ENVIRONMENTAL INDICATORS AND AGRICULTURAL POLICY

Environmental Indicators and Agricultural Policy

Edited by

Floor Brouwer
Agricultural Economics Research Institute, The Hague, The Netherlands

and

Bob Crabtree
Macaulay Land Use Research Institute, Aberdeen, UK

CABI *Publishing*

CABI *Publishing*
CAB INTERNATIONAL
Wallingford
Oxon OX10 8DE
UK

CABI *Publishing*
10 E 40th Street
Suite 3203
New York, NY 10016
USA

Tel: +44 (0) 1491 832111
Fax: +44 (0) 1491 833508
Email: cabi@cabi.org

Tel: +1 212 481 7018
Fax: +1 212 686 7993
Email: cabi-nao@cabi.org

A catalogue record for this book is available from the British Library, London, UK.

Library of Congress Cataloging-in-Publication Data
Environmental indicators and agricultural policy/edited by
  Floor Brouwer and Bob Crabtree.
      p.   cm.
    Includes bibliographical references and index.
    ISBN 0-85199-289-7
    1. Agriculture--Environmental aspects--European Union countries.
  2. Agriculture and state--Environmental aspects--European Union
  countries.   3. Environmental indicators--European Union countries.
  I. Brouwer, Floor.   II. Crabtree, Bob.
  S589.76.E85E58   1998
  333.76'14--dc21   98-27951
                                                                          CIP

ISBN 0 85199 289 7

Typeset in Photina by AMA Graphics Ltd, UK
Printed and bound in the UK by Biddles Ltd, Guildford and King's Lynn

Contents

Part II: Indicators in the Field of Biodiversity and Landscape

Part III: Indicators in the Field of Environmental Pollution

Part IV: Indicators in the Field of Policy Analysis

Contributors

David Baldock is Director of the Institute for European Environmental Policy, Dean Bradley House, 52 Horseferry Road, London SW1P 2AG, UK.

Floor M. Brouwer is head of the research unit, Environment and Technology, Agricultural Economics Research Institute, PO Box 29703, 2502 LS The Hague, The Netherlands.

Jan Buys is a project leader in agriculture, nature and water management at the Centre for Agriculture and Environment, PO Box 10015, 3505 AA Utrecht, The Netherlands.

Bob Crabtree is Head of the Environmental and Socio-Economics Group at the Macaulay Land Use Research Institute, Craigiebuckler, Aberdeen AB15 8QH, UK.

Stephan Dabbert is Professor of Agricultural and Resource Economics, Institut für Landwirtschaftliche Betriebslehre (410A), Universität Hohenheim, D-70593 Stuttgart, Germany.

Juha Grönroos is a research scientist, Finnish Environment Institute, Pollution Prevention Division, PO Box 140, FIN-00251, Helsinki, Finland.

Hubert Gulinck is professor in the field of landscape ecology, Catholic University of Leuven in Belgium, and member of the International Association of Landscape Ecologists, European Centre for Nature Conservation (ECNC), PO Box 1352, 5004 BJ Tilburg, The Netherlands.

Jochen Jesinghaus coordinates the European Commission's Environmental Pressure Indices Programme, Commission of the European Communities, DG 34/F3, JMO C4/007, L-2920 Luxembourg.

Bernard Kilian is Research Associate, Institut für Landwirtschaftliche
    Betriebslehre (410A), Universität Hohenheim, D-70593 Stuttgart,
    Germany.
Philip Lowe is Duke of Northumberland Professor of Rural Economy and
    Director of the Centre for Rural Economy at the University of Newcastle
    upon Tyne, Newcastle upon Tyne NE1 7RU, UK.
Juan E. Malo is Senior Researcher in Ecology and Natural Resources,
    Departamento Interuniversitario de Ecología, Universidad Autonoma
    de Madrid, 28049 Madrid, Spain.
Markus Meudt is a research assistant at the Institute for Agricultural Policy,
    Market Research and Economic Sociology, Nussallee 21, D-53115
    Bonn, Germany. (*Present address*: Orchideenweg 23, D-53123 Bonn,
    Germany.)
Marta Múgica is Project Manager at the Centro des Investigaciones
    Ambientales de la Comunidad de Madrid and guest lecturer at the
    Complutense University, Madrid in Spain and also at European Centre
    for Nature Conservation (ECNC), PO Box 1352, 5004 BJ Tilburg, The
    Netherlands.
Juan J. Oñate is Lecturer in Natural Resources and Environmental Impact,
    Departamento Interuniversitario de Ecología, Universidad Autonoma
    de Madrid, 28049 Madrid, Spain.
Arie J. Oskam is Professor of Agricultural Economics and Policy at the
    Department of Economics and Management, Wageningen Agricultural
    University, Hollandseweg 1, 6706 KN Wageningen, The Netherlands.
    (Visiting Professor at the Department of Agricultural and Resource
    Economics, University of Maryland.)
Kevin Parris is a principal economist, Environment Division, Directorate for
    Food, Agriculture, and Fisheries, OECD, 2 Rue André Pascal, 75775
    Paris, France.
Begoña Peco is Professor of Ecology and System Analysis and Modelling,
    Departamento Interuniversitario de Ecología, Universidad Autonoma
    de Madrid, 28049 Madrid, Spain.
Reijo Pirttijärvi is Senior Advisor in the Ministry of Agriculture and
    Forestry, The Department of Agriculture, PO Box 232, FIN-00411,
    Helsinki, Finland.
Clive Potter is Lecturer in Environmental Policy in the Environment
    Department, Wye College, University of London, Wye, Ashford, Kent
    TN25 5AH, UK.
Ewald Rametsteiner is Scientific Assistant at the Institute of Forest Sector
    Policy and Economics at the Universität für Bodenkultur, Gregor
    Mendel Str. 33, 1180 Vienna, Austria.
Seppo Rekolainen is a senior scientist, Finnish Environment Institute,
    Impacts Research Division, PO Box 140, FIN-00251, Helsinki, Finland.
Eirik Romstad is a senior research fellow in environmental economics and
    head of research at the Department of Economics and Social Sciences,

Agricultural University of Norway, PO Box 5033, N-1432 Aas, Norway.

Sabine Sprenger is Research Associate, Institut für Landwirtschaftliche Betriebslehre (410A), Universität Hohenheim, D-70593 Stuttgart, Germany.

Francisco Suárez is Senior Lecturer of Natural Resources Environmental Impact and Conservation Biology, Departamento Interuniversitario de Ecología, Universidad Autonoma de Madrid, 28049 Madrid, Spain.

José Sumpsi is Professor of Agricultural Economics, Departamento de Economía y Ciencias sociales Agrarias, Universidad Politecnica de Madrid, 28040 Madrid, Spain.

Graham Tucker is a senior consultant with Ecoscope Applied Ecologists, 9 Bennell Court, Comberton, Cambridge, CB3 7DS, UK.

Rob Vijftigschild is Researcher of the Agricultural Economics and Policy Group, Department of Economics and Management, Wageningen Agricultural University, Hollandseweg 1, 6706 KN Wageningen, The Netherlands.

Neil Ward is a lecturer in the Department of Geography, University of Newcastle upon Tyne, Newcastle upon Tyne NE1 7RU, UK.

Dirk M. Wascher is Senior Programme Coordinator for Biodiversity and Landscapes at the European Centre for Nature Conservation in Tilburg, The Netherlands.

Jaap van Wenum is a doctoral student at the Department of Economics and Management, Wageningen Agricultural University, PO Box 8130, 6700 EW Wageningen, The Netherlands.

Ada Wossink is Associate Professor at the Department of Economics and Management, Wageningen Agricultural University, PO Box 8130, 6700 EW Wageningen, The Netherlands.

Preface

The present book on environmental indicators and agricultural policy includes edited and revised versions of papers presented at the Workshop 'Towards operationalization of the effects of CAP on environment, landscape and nature: Exploration of indicator needs', held in Wageningen in April 1997. The workshop was organized by the Agricultural Economics Research Institute (LEI-DLO) in The Netherlands. One of the prime objectives of the workshop was to integrate results of the four workshops on pesticides, minerals, global warming and landscape and nature, which had been organized within the Concerted Action AIR3 - CT93 - 1164 'Policy measures to control environmental impacts from agriculture'. The workshop focused on practical methods for translating research results into environmental indicators for the agricultural sector in the EU.

ACKNOWLEDGEMENTS

The editors would like to thank the following persons for their advice in selecting and revising first drafts of the papers: Paul Berentsen, Wageningen Agricultural University, The Netherlands; Henry Buller, Université de Paris, France; Alex Dubgaard, Royal Veterinary and Agricultural University, Denmark; Ian Hodge, University of Cambridge, United Kingdom; Markus Hofreither, University of Bodenkultur, Vienna, Austria; Guido van Huylenbroeck, University of Gent, Belgium; Onno Kuik, Institute for Environmental Studies, Amsterdam, The Netherlands; Michael Linddal, Danish Institute of Agricultural and Fisheries Economics, Copenhagen, Denmark; Asko Miettinen, Agricultural Economics Research Institute, Helsinki, Finland;

Andrea Povellato, National Institute of Agricultural Economics, Padova, Italy; John Sumelius, University of Helsinki, Finland; Martin Whitby, Centre for Rural Economy, University of Newcastle upon Tyne, United Kingdom; Ada Wossink, Wageningen Agricultural University, The Netherlands.

Editorial assistance was provided by Freda Miller, Centre for Agricultural Strategy, University of Reading. She guided the communication process with the authors and took responsibility for carefully cross-checking the text. We also acknowledge secretarial assistance provided by Carol Smith, Macaulay Land Use Research Institute, Aberdeen.

The editors wish to express their appreciation to Arie Oskam, coordinator of the Concerted Action, for the arrangements he made to support the organization of the workshop and for his encouragement and support in preparing the text for publication.

Floor Brouwer
Bob Crabtree

Introduction

Floor Brouwer and Bob Crabtree

1

INDICATORS IN ECONOMIC POLICY

Indicators are widely used in the evaluation of sectoral policy to judge the state of national economies and individual sectors. Gross domestic product (GDP), for example, is a very important measure of economic output over a given period of time. It provides a simultaneous measure of economic output generated by production and the goods and services produced and available for consumption, investment and export. This indicator has a long history in policy evaluation, both nationally and internationally, and presently is developed along harmonized procedures and is relatively simple to measure. It shows changes over time and allows for cross-national comparisons.

Single-issue indicators have also been developed for use in agricultural policy evaluation. Production-linked support to the agricultural sector is an important indicator of government intervention in the sector. The producer subsidy equivalent (PSE) is an important measure of the extent of agricultural subsidy, which distinguishes between market and price support, direct payments and other support. It is a single indicator of the monetary transfers from taxpayers (through government budgets) and consumers (through domestic price support at levels that exceed world market prices) to agricultural production. PSEs are calculated on a periodic basis for most of the OECD countries, allowing for trends over time and cross-national comparisons.

An indicator – like the ones presented here – is defined as a 'parameter, or a value derived from parameters, which points to, provides information about, describes the state of a phenomenon/environment/area, with a significance extending beyond that directly associated with a parameter value' (OECD, 1994). Two important features of indicators are quantification of information as well as simplification of complex phenomena (Hammond *et al.*, 1995). Such notions are important to allow simplification of the communication process among users, but also to reduce the number of measurements and parameters required to present a situation. The use of consistent methodologies is fundamental in order to allow for cross-national and long-term comparisons of such an indicator.

Relationships between agriculture and the environment are relatively complex because of the diversity of production systems, with wide geographical heterogeneity and large temporal variability in European agriculture, and a wide range of biophysical conditions across Europe. Agri-environmental linkages also are characterized by the vulnerability and unpredictability of ecosystems. Given such non-straightforward links between agriculture and the physical environment, landscape and nature, a systematic approach with high flexibility is needed. In order to achieve this, the formulation of agri-environmental indicators could provide helpful tools to facilitate monitoring of agricultural and environmental policies.

The reform of agricultural policy in the European Union (EU) and the formulation of environmental policy objectives at EU level increasingly require well-defined and targeted environmental measures. Methods are therefore needed to indicate the effectiveness of policy response through the agri-environment programmes. They are also required with the looming World Trade Organisation (WTO) round and the need to justify agricultural support in terms of environmental benefits. Indicators therefore are required to judge whether a reduction in production-linked agricultural support would be beneficial or harmful to the environment. From a policy perspective, the development of agri-environmental indicators is important in the context of agricultural policy reform. Such indicators will measure the impacts of agricultural policy reform in the European Union on the physical environment, landscape and nature and contribute to the achievement of policy targets.

ENVIRONMENTAL CONCERNS AND AGRICULTURE

There is an increasing need to assess the environmental impacts of societal trends. The European Environment Agency (EEA), for example, responds through periodic reports on the state of the environment (Wieringa, 1995). The impacts of societal trends (industry, energy, transport, tourism, agriculture) have been explored and elaborated by environmental theme. The themes are distinguished at a global scale (e.g. climate change), the European and transboundary scale (e.g. acidification) and at regional scale (e.g. waste management).

Linkages between agriculture and the environment are far more complex than with other economic activities. The impact of agriculture on the environment includes beneficial and benign effects and distinguishes between:

- Soil quality (erosion, nutrient supply, moisture balance). Losses of nitrogen from agriculture, for example, include leaching of nitrates to surface water and groundwater, as well as denitrification. High levels of animal manure may create problems with soil pollution.
- Water quality (leaching of nutrients and pesticides, water extraction and drainage). There is some empirical evidence that the use of pesticides poses a threat to the environment; also, nitrate levels in groundwater need to be reduced and eutrophication of surface water is a particular problem in certain parts of the European Union.
- Air quality (emissions of ammonia and greenhouse gases). Emissions of ammonia, of which agriculture is a major source, contribute to the acidification of soils and water. Greenhouse gas emissions have the potential to alter global climatic conditions.
- Biodiversity, including the variability among all living organisms. Low-intensity livestock production has created large areas of semi-natural grassland, scrubland, heather moorland and other grazed habitats of value in nature conservation.
- Landscape, including preservation of landscapes by farming systems with high nature value. Animal production sectors, such as beef, sheep and dairy farming, are important for the provision of public goods in the field of rural environment and landscapes in large areas of Europe, as they manage the areas of high nature conservation value. They are often regarded as low-intensity farming systems with a wide diversity of habitat systems, although relatively intensive grazing systems, such as the peat areas in parts of The Netherlands, may also have high nature conservation potential (see also, for example, Brouwer and Van Berkum, 1996; OECD, 1998).

Knowledge of the environmental effects of agricultural policy is required in order to ensure that 'environmental concerns are taken into account from the outset in the development of policies and in the implementation of those policies, and the need of appropriate mechanisms within the Member States', as stated by the Council of Ministers and the Representatives of the Governments of Member States' Meeting with the Council on the Fifth Environmental Action Programme. Indicators would be useful in giving a broad picture of development to try to better understand the relationships between policy, agricultural production and the environment.

MONITORING AGRICULTURAL AND ENVIRONMENTAL POLICY

The increasing notion of linkages between agriculture and the physical environment, as well as with landscape and biodiversity, required monitoring efforts on progress to:

1. support the evaluation of environmental policy objectives related to the agricultural sector;

2. examine the response by the agricultural sector towards meeting standards on quality of the physical environment, biodiversity and landscape; and
3. investigate options available to the agricultural sector in meeting environmental targets. Present farming practice, for example, varies widely in its use of agrochemicals across homogeneous groups of holdings.

The development of environmental indicators in the context of agricultural policy (OECD, 1997) aims to:

1. provide information to policy-makers and society about the current state of the environment and its evolution over time;
2. assist policy-makers in improving their understanding of linkages between agriculture and the environment; and
3. contribute to monitoring and evaluation efforts of policies for the achievement of environment-friendly production methods.

Agri-environmental indicators are important in the assessment of trends over time of the effects of agriculture on the environment. The contribution of the Common Agricultural Policy (CAP) towards the achievement of environmentally sustainable forms of production was an important part of the reform of CAP in 1992. Sustainable development was defined in the Brundtland Report as 'development which meets the needs of the present without compromising the ability of future generations to meet their own needs'. This notion integrates both economic, environmental and social factors. It is also given more emphasis in the context of European policy by the formulation of Article 130R to integrate the requirements of the environment into other Community policies.

The United Nations Commission on Sustainable Development (CSD) initiated efforts to create sustainable development indicators. It has requested countries to use indicators in their attempts to measure progress in achieving sustainable development, according to Agenda 21 adopted at the UNCED Rio Summit in 1992. Agenda 21 comments specifically on the need for indicators in Chapter 40, where reference is made to 'Indicators of sustainable development need to be developed to provide solid bases for decision making at all levels and to contribute to self-regulatory sustainability of integrated environment and development systems'. The indicators are organized in the driving force–state–response framework (DSR). In the framework of such a concept 'Driving Forces represent human activities, processes and patterns that impact on sustainable development, State indicators indicate the "state" of sustainable development, and Response indicators indicate policy options and other responses to changes in the state of sustainable development' (United Nations, 1996). Indicators of sustainable development are proposed for the categories social, economic, environmental and institutional (United Nations, 1996; Table 1.1). Methodologies to establish the sustainable development indicators are presented in that report. Following the methodologies proposed by the CSD, a pilot study was initiated by Eurostat (Eurostat, 1997). The Framework of driving forces–state–response is taken up by Eurostat, OECD and the EEA in their efforts to integrate driving forces to the state of the environment.

THE ORGANIZATION OF THE BOOK

The book has five parts: (I) introduction; (II) indicators in the field of bio-diversity and landscape; (III) indicators in the field of environmental pollution; (IV) indicators in the field of policy analysis; and (V) conclusions. Each part includes chapters that focus on practical approaches to translating the research results on environmental indicators for the agricultural sector in the European Union.

The first part of the book is an introduction to environmental indicators and agricultural policy, reviewing efforts made in recent years to assess the impacts of agricultural policy on the physical environment, landscape and biodiversity. Progress has been achieved mainly by a proper input from the scientific and policy communities.

The book starts with some theoretical considerations, provided by Eirik Romstad, in the development of environmental indicators. Criteria for the selection of indicators include consistency, reliability, predictive capacity and benefit/cost measures. Indicators may contribute to reducing the marginal costs of data gathering efforts. The optimal amount of information to collect is where the expected marginal cost of obtaining additional information equals its expected marginal benefits.

The next chapter provides an overview of OECD work on agri-environmental indicators. Kevin Parris emphasizes the need to evaluate the impact of policy measures, both harmful and beneficial, to the environment, landscape and biodiversity. Indicators should guide governments and other users in their efforts to improve targeting of agricultural and environmental programmes and to monitor and assess policies. Methods of measurement are established for some indicators (e.g. greenhouse gases and nutrient balances); the conceptual and analytical understanding of some indicators require effort, in particular biodiversity, landscape and socio-cultural aspects of agri-environmental linkages.

Jochen Jesinghaus, in his contribution on agricultural sector pressure indicators in the European Union (EU), tries to build the bridge between environmental scientists and decision-makers. The overall objective of Chapter 4 is to produce a tool that serves environmental policy in a way similar to that of the system of national accounts which has served economic policy over the past decades. Indicators are identified for the policy field Loss of Biodiversity. Since the production of policy-relevant indicators is a time-consuming and costly exercise, he proposes means to identify and rank indicators.

Chapter 5 reviews indicator requirements from a policy point of view, and matches requirements with data availability. Agri-environmental indicators are required increasingly to serve both monitoring of environmental

Table 1.1. List of environmental indicators of sustainable development (source: United Nations, 1996, pp. x–xii).

Chapter of Agenda 21	Driving force indicators	State indicators	Response indicators
18 Protection of the quality and supply of freshwater resources	Annual withdrawals of ground and surface waters Domestic consumption of water per capita	Groundwater reserves Concentration of faecal coliform in fresh water Biochemical oxygen demand in water bodies	Waste-water treatment coverage Density of hydrological networks
17 Protection of the oceans, all kinds of seas and coastal areas	Population growth in coastal areas Discharges of oil into coastal waters Releases of nitrogen and phosphorus to coastal waters	Maximum sustainable yield for fisheries Algae index	
10 Integrated approach to the planning and management of land resources	Land-use change	Changes in land conditions	Decentralized local-level natural resource management
12 Managing fragile ecosystems: combating desertification and drought	Population living below poverty line in dryland areas	National monthly rainfall index Satellite-derived vegetation index Land affected by desertification	
13 Managing fragile ecosystems: sustainable mountain development	Population change in mountain areas	Sustainable use of natural resources in mountain areas Welfare of mountain populations	
14 Promoting sustainable agriculture and rural development	Use of agricultural pesticides Use of fertilizers Irrigation per cent of arable land Energy use in agriculture	Arable land per capita Area affected by salinization and waterlogging	Agricultural education
11 Combating deforestation	Wood harvesting intensity	Forest area change	Managed forest area ratio Protected forest area as a percentage of total forest area

Table 1.1.   *contd.*

Chapter of Agenda 21	Driving force indicators	State indicators	Response indicators
15 Conservation of biological diversity		Threatened species as a percentage of total native species	Protected area as a percentage of total area
16 Environmentally sound management of biotechnology			R&D expenditure for biotechnology Existence of national biosafety regulations or guidelines
9 Protection of the atmosphere	Emissions of greenhouse gases Emissions of sulphur dioxide Emissions of nitrogen oxides Consumption of ozone depleting substances	Ambient concentration of pollutants in urban areas	Expenditure on air pollution abatement
21 Environmentally sound management of solid wastes and sewage-related issues	Generation of industrial and municipal solid waste		Expenditure on waste management Waste recycling and reuse Municipal waste disposal
19 Environmentally sound management of toxic chemicals		Chemically induced acute poisonings	Number of chemicals banned or severely restricted
20 Environmentally sound management of hazardous wastes	Generation of hazardous wastes Imports and exports of hazardous wastes	Area of land contaminated by hazardous wastes	Expenditure on hazardous waste treatment
22 Safe and environmentally sound management of radioactive wastes	Generation of radioactive wastes		

legislation in the EU as well as responses by the agricultural sector to improve environmental soundness of agricultural production. The various recent initiatives to develop agri-environmental indicators, both at international and national level, are reviewed by Floor Brouwer.

The second part of the book is on indicators in the field of biodiversity and landscape. The conceptual and analytical understanding of the links between agriculture and indicators for such areas still require major efforts.

In Chapter 6 targets are established to assess agricultural impacts on European landscapes. Dirk Wascher and his co-authors develop a framework concept for landscape types. A typology of European landscapes and standard methodologies allow landscape types to be characterized in a comparable manner. A number of conceptual and qualitative challenges with regard to data management are pointed out. These are specifically relevant to the use of indicators or the development of implementation targets in particular. Procedures are required to link top-down approaches by international experts with national bottom-up activities. Also, an active dialogue is essential across several disciplines, including natural and social scientists, as well as policy analysts.

Chapter 7 reviews the potential role and limitations of using indicators to measure the impacts of agriculture on biodiversity. General properties of good indicators are examined. Based on these properties, suitable indicators are identified by Graham Tucker, for measuring pressures on biodiversity, the state of biodiversity and responses to biodiversity losses, and examples are discussed. Lastly, proposals are made for further data requirements for monitoring agricultural impacts on biodiversity.

Criteria for the application of nature quality indicators in agriculture are developed in Chapter 8. Some recent attempts to design nature quality indicators are reflected on by Jaap van Wenum, Jan Buys and Ada Wossink. In addition, the authors propose a yardstick for biodiversity as an instrument to quantify and judge biodiversity on individual holdings. In order to serve analyses on the effects of nature conservation measures, the precision of the yardstick can be improved through the provision of detailed monitoring data on various species.

Chapter 9 is on indicators for high nature value (HNV) farming systems in Europe. David Baldock elaborates on the 'rediscovery' in policy analysis of HNV farming. Three approaches are identified, and indicators are defined for each of them. Indicators are defined for environmentally protected areas, based on species and habitat management requirements, and for the identification of low-intensity farming systems. Pragmatic approaches of using existing information sources may be preferable for defining HNV farming systems more precisely.

Indicators for extensive land-use systems in the Iberian Peninsula are developed by Begoña Peco and her colleagues in Chapter 10. Indicators are defined for dry cereal steppes and *dehesas*. Most of the indicators proposed reflect driving forces rather than acting as indicators on the state of the environment. Statistical data are available to a limited extent only, and options are provided to assess such farming systems at a regional, or even at a local or farm level. The use of geographical information systems may overcome many of the limitations of the availability of statistical data.

A state-of-the art review of indicators in the field of environmental pollution is provided in the third part of the book. Emphasis is given to water-quality indicators related to nutrient balance and pesticide use.

Environmental pressure indicators for pesticide impacts are described in Chapter 11. Arie Oskam and Rob Vijftigschild review environmental indicators used so far in the policy field of pesticides. It is argued that risk aspects and the quantity of pesticides should both play a role in the identification of indicators. A principal component analysis is applied to derive a weighted indicator from several independent basic indicators.

Chapter 12 develops a system of site-specific water-quality indicators in Germany. Stephan Dabbert and his colleagues demonstrate the use of such a system to support extension agents to evaluate the impact of farming practices on groundwater and surface water. The transaction costs (e.g. costs for the provision of information, monitoring and administration) also need to be considered to judge the cost-effectiveness of agri-environmental indicators. A distinction is made between water-quality indicators, which have a high validity but are difficult to obtain, and indicators that are relatively easy to obtain but may have a much lower validity. For extension purposes the type of indicators used needs to be close to the issue of concern (e.g. water-quality problems).

Chapter 13 covers the use of nutrient balances in the implementation of agricultural policy measures in Finland. Reijo Pirttijärvi and co-workers compare two approaches to the calculation of nutrient balances, and reflect on some critical driving forces. Weather variations, for example, may contribute to substantial interannual variation in nutrient surpluses. Nutrient balances are also being used in the framework of the agri-environmental measures implemented in that country.

Environmental indicators are applied in the field of policy analysis. Part 4 of the book reviews attempts made on the use of environmental indicators for policy analysis.

Bob Crabtree, in Chapter 14, examines the potential of sustainability indicators and the extent to which they can be identified for multiple land use. The work is based on an empirical analysis for an environmentally sensitive area in the uplands of the United Kingdom. At local level the scope for developing sustainability indicators is limited by transboundary effects and differences in the spatial scales of the environmental dimensions. Their role is to inform local stakeholders on economy–environment interactions located in local space. Sustainability indicators in general are as much shaped by the policy process as developed to inform the process.

The implementation of environmental indicators in policy information systems is discussed by Markus Meudt. A detailed regional modelling system is used in Chapter 15 as a tool for integrated agricultural and environmental policy analysis. Indicators used include nutrient balances, biodiversity and a land-use diversity indicator, and emissions of greenhouse gases. The modular approach and flexible adjustments allow for sufficient regional detail and the incorporation of environmental indicators.

Chapter 16 provides a critical review on the criteria and indicators developed in the forestry sector. Ewald Rametsteiner supports the hypothesis that recent increased efforts in developing indicators in forestry have their roots in the political goals of striving towards sustainable forestry management. Ecological, economical and social aspects are the main dimensions in such a complex normative abstract and dynamic concept. A hierarchical system of principles, criteria and indicators has been developed to cover such aspects in policy formulation and evaluation. The main practical application of indicators is in the evaluation of and reporting on the achievement of national forest policy to ensure sustainable forest management, and in the use as assessment tools in the context of certification of sustainable forest management.

Chapter 17 is on attitudinal and instititutional indicators for sustainable agriculture. Philip Lowe, Neil Ward and Clive Potter address such indicators to explore whether policy is set in an appropriate direction. This might be appropriate since the full environmental consequences of policy measures taken in recent years may not become clear for several years to come. Also, this type of indicator of altered social relationships and institutional structures would be essential in a full appreciation of permanently sustainable socio-economic systems, with an effective integration of environmental objectives into agriculture.

The final part of the book reviews critically the present state of understanding on using environmental indicators in the evaluation of agricultural policy. Bob Crabtree and Floor Brouwer judge current knowledge and provide an outlook towards future efforts to further improve our knowledge regarding linkages between agricultural policy, the physical environment, landscape and biodiversity.

REFERENCES

Brouwer, F.M. and Van Berkum, S. (1996) *CAP and Environment in the European Union: Assessment of the Effects of the CAP on the Environment and Assessment of Existing Conditions in Policy*. Wageningen Pers, Wageningen, The Netherlands.

Eurostat (1997) *Indicators of Sustainable Development: a Pilot Study Following the Methodology of the United Nations Commission on Sustainable Development*. Statistical Office of the European Communities (Eurostat), Luxembourg.

Hammond, A., Adriaanse, A., Rodenburg, E., Bryant, D. and Woodward, R. (1995) *Environmental Indicators: A Systematic Approach to Measuring and Reporting on Environmental Policy Performance in the Context of Sustainable Development*. World Resources Institute, Washington.

OECD (1994) *Environmental Indicators: OECD Core Set*. Organisation for Economic Co-operation and Development, Paris.

OECD (1997) *Environmental Indicators for Agriculture*. Organisation for Economic Co-operation and Development, Paris.

OECD (1998) *The Environmental Effects of Reforming Agricultural Policy*. Organisation for Economic Co-operation and Development, Paris.

United Nations (1996) *Indicators of Sustainable Development Framework and Methodologies*. United Nations, Division for Sustainable Development, Department for Policy Coordination and Sustainable Development, New York.

Wieringa, K. (ed.) (1995) *Environment in the European Union 1995: Report for the Review of the Fifth Environmental Action Programme*. European Environment Agency, Copenhagen, Denmark.

Theoretical Considerations in the Development of Environmental Indicators[1]

Eirik Romstad

INTRODUCTION

Indicators are used because they are perceived as a low-cost way of improving the basis for decision-making in contexts where information is incomplete. This implies that indicators rarely provide a complete description of a particular problem, and that they should not be used in isolation from other pertinent information available to the decision-maker(s).

Norgaard (1990) stated that economic indicators provide little valuable information on resource scarcity, as an economy consisting of uninformed agents would not result in prices reflecting this scarcity. In brief, Norgaard argued that economic indicators only measure the beliefs of agents about scarcity. Farrow and Krautkraemer (1991), amongst others, challenged Norgaard's findings, arguing that resource allocators have incentives to be well informed, and consequently that economic indicators will provide better information than physical indicators. Brekke (1994) and Asheim (1994) both argued that, if coupled with sufficient auxiliary assumptions, economic indicators – like net national product – provide reliable welfare information.

In the case of pollution, for example from agriculture, the use of economic indicators is more problematic. First, pollution is an externality, implying that its effects on the profits of the polluters are minor unless the externality is internalized by some policy. Second, valuation of less polluted water, for example, through contingent valuation[2] (or some other valuation method), gives a measure of the scarcity of clean water, not of the effects of certain policies for the improvement of water quality. For the latter purpose, physical indicators may be more appropriate in many cases.

©CAB INTERNATIONAL 1999. *Environmental Indicators and Agricultural Policy* (eds F.M. Brouwer and J.R. Crabtree)

Environmental problems are multifaceted and often complex. One implication of this is that it is virtually impossible and costly to measure all environmental aspects. The need for an ordered way of collecting information on the state of the environment is therefore self-evident. An environmental indicator can be viewed as a targeted way of collecting information on the state of the environment.

Targeting in incomplete information settings raises several questions. One is the ability of the indicator(s) to capture the state (or changes in the state) of the environment, i.e. the informational quality of the indicator(s). Another aspect is that of cost, which raises an additional main criterion: benefits/cost considerations. The expected value of the information gained by the use of some indicator(s) must be compared to the cost of collecting and analysing the necessary data. By applying theoretical concepts such as the *expected value of perfect information* (Bunn, 1984), such benefits/cost considerations can be undertaken in a systematic way.

Although quality aspects and costs should be considered simultaneously when deciding on which indicator(s) to use, treating them separately makes it easier to conduct an orderly discussion, highlighting their single and joint importance for decision-making under uncertainty.

The remainder of this chapter covers the quality aspects of indicators in more detail, the questions related to the optimal amount of information to be collected and processed, and the applicability of physical environmental indicators in agriculture, exemplified by farm-gate nutrient balances and ambient water quality in the recipient(s).

THE QUALITY OF AN INDICATOR

The quality aspects of an indicator can be divided into three elements: (i) consistency, (ii) reliability and (iii) predictive capacity. The Organisation for Economic Cooperation and Development (OECD) uses other quality criteria: *measurability, policy relevance* and *analytical soundness*. Indicators are chosen primarily because they are easy to measure.

There are two major reasons why policy relevance is not considered here as a quality criterion:

1. Indicators are often designed to capture changes that occur over time. Often, this requires long-term time-series data. Collecting high-quality time-series data is a time-consuming process. Limiting current indicators to those relevant to current policies may therefore conflict with the precautionary principle.

2. Even though a close link between an environmental indicator and policy is preferable, one should be careful with such additional criteria. The main reasons for this are much the same as those used in the literature on regulation: incentive compatibility and Pareto-optimality are often jointly not attainable

(Hurwicz, 1972; Campbell, 1987). The analogous argument for indicators is that policy relevance may come at the cost of quality performance of the indicators.

The final criterion listed by the OECD is analytical soundness. Contrast this criterion with the following three quality aspects of indicators:

1. Consistency: the indicator must capture changes in key state variables in a way that is comparable over time. For example, using biological or chemical agents found in ecosystems as indicators may cause difficulties, as these variables may be subject to stochastic processes. Weather is a prime example of a collection of such processes.

2. Reliability: many environmental problems evolve over time. Hence, long-term quality time-series data with well-suited resolution level(s) are prime candidates on which to base a reliable indicator, from which reliable inferences can be drawn.

3. Predictive capacity: variables that display certain characteristics before the environmental problem in question becomes large (leading indicators) are preferable as indicators. The ability of the indicator(s) to identify risk and capture changes over time are of particular importance in this connection.

Together, these three criteria imply analytical soundness, but analytical soundness does not necessarily imply that all these three criteria are met. To see this, and at the same time to see how the three listed criteria are connected, consider biological oxygen demand (BOD), a frequently used indicator. Consistency, reliability and predictive capacities are all improved if coupled to models and experimental trials. For example, measuring BOD in waterways without the ability to adjust ('whiten') these measurements for variations in water temperature, water flow, etc., may reduce the informational value of BOD measurements.

THE OPTIMAL AMOUNT OF INFORMATION COLLECTION UNDER RISK AND UNCERTAINTY

Gathering information on the state of the environment is costly. Better targeting of this information collection process is one way to reduce these costs. Indicators can be viewed as highly targeted information. The cost of monitoring the environment is, however, only one aspect. There are also clear benefits from having good data on the state of the environment. Early detection of environmental degradation makes it easier and cheaper to take pre-emptive measures, and increases the likelihood of achieving higher environmental quality.

The literature on the economics of information is clear in its recommendations: the optimal amount of information collection is where the expected marginal costs of gathering information equal the expected benefits of obtaining this information (see, for example, Bunn, 1984). Obtaining high-quality information on the state of the environment facilitates better decision-making with respect to environmental problems.

In an incomplete information setting, indicators help the decision process in several ways. Two aspects stand out in particular:

- In deciding on what indicators to use, decision-makers are 'forced' to assess the expected benefits of the information contained in the indicator.
- The marginal cost of collecting information is clarified.

These two effects influence the optimal level of information to gather (Ω), as depicted in Fig. 2.1.

Besides showing the optimal level of information to collect, Ω, Fig. 2.1 depicts two important and partly related aspects of information collection:

1. The width of the confidence intervals tends to decrease as more information is collected, i.e. the uncertainties are reduced. For risk-averse decision-makers or when damage functions are convex, increased precision (reduced uncertainty) is important by itself (Romstad and Vatn, 1995).

2. The larger the uncertainty (wider confidence intervals), the more uncertain is the optimal information level (the distance between Ω_L and Ω_H increases). Viewing the upper confidence interval for the marginal costs and the lower confidence interval for the marginal benefits as the true estimates for marginal costs and benefits, the figure also shows that the optimal amount of information declines to Ω_L. Conversely, the higher the marginal benefits and the lower the marginal costs, the more the optimal amount of information increases.

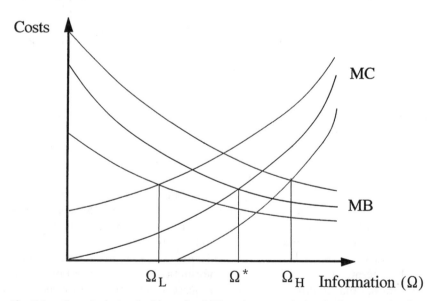

Fig. 2.1. Expected marginal benefits (MB) and expected marginal costs (MC) of information. Bordering lines depict confidence intervals around the expected values.

In some ways Fig. 2.1 is an oversimplification. First, information, Ω, may not be continuous, but rather discrete, linked to the various indicators. This discreteness would imply that the marginal benefit and cost curves depicted in the figure could become stepped. Even so, the optimality principle of equating marginal benefits and costs, holds. Secondly, using several indicators together could result in negative covariances between indicators, reducing the uncertainty of certain indicator combinations. When choosing a set of indicators, this possibility should be considered as it reduces the uncertainty in the optimal amount of information to collect.

The marginal benefits of collecting additional information also include two other principally different, but related issues: (i) the welfare implications, and (ii) the environmental impacts of (continued) pollution. In terms of Fig. 2.1, large negative welfare impacts (for example, established through valuation studies) or environmental effects manifest themselves in an upward shift of the marginal benefit curve. Consequently the optimal level of information to collect increases.[3]

PHYSICAL ENVIRONMENTAL INDICATORS AND AGRICULTURAL POLLUTION

As mentioned in the introduction, the choice of an indicator or a set of indicators depends on two main aspects, what may be termed loosely 'quality' (i.e. a combination of consistency, reliability and predictive capacity) and costs. Assume policy-makers' preference for indicators can be explained as a trade-off between 'quality' and costs, such that an increase in 'quality' can be offset by an increase in costs at a decreasing rate. This implies diminishing marginal utility for 'quality', which is consistent with standard assumptions in cardinal utility theory. In a 'quality'–cost indifference diagram this suggests sloping curves from the origin or from the quality axis to the north-east, as shown in Fig. 2.2.[4]

The two indifference mappings, U_i^q and U_i^c, denote preferences with relatively more weight put on 'quality' and cost aspects, respectively, while the bubbles marked (a–d) denote four discretely different indicators. From Fig. 2.2 it is easy to see that decision-maker preferences could be the factor determining which indicator (set) to use. When 'quality' aspects are relatively more important (preferences U_i^q), indicator (d) is the preferred choice (indifference curve U_2^q goes through (d) and gives higher utility than any of the three other indicators). Conversely, when cost considerations are relatively more important, indicator (b) becomes an interesting option (indifference curve U_2^c goes through (b) and gives higher utility than any of the three other indicators). Also note that indicator (c) will never be chosen as it lies to the north-west of (d).

The ellipses around the four indicators (a–d) indicate the 'quality'–cost uncertainties associated with each indicator. Generally, one would expect that

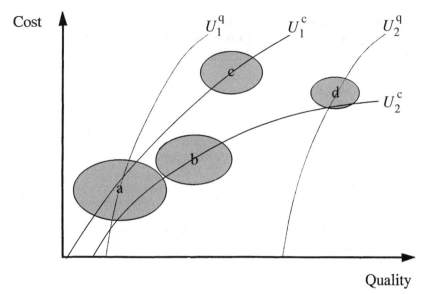

Fig. 2.2. Ranking of expected indicator quality and expected costs (U^c and U^q denote different mappings of cost–accuracy indifference curves; subscript i (= 1 or 2) denotes preference ranking within each indifference curve mapping; a–d denote indicators).

it is more difficult to assess the 'quality' of an indicator relative to its operating costs as depicted by the ellipses in Fig. 2.2. With indicator (d) having the least uncertainty (smallest ellipse area), relative large weight on 'quality' would increase the likelihood of (d) being the preferred indicator.

Cost- or Quality-weighted Preferences

The applicability of an indicator depends on the site-specific environmental state. The capacity of an ecosystem to sustain emissions is likely to vary depending on its self-cleaning capacity. It is well known that the past history of emissions to a system influences its self-cleaning capacity (compare with literature on stock pollutants: Conrad, 1992). The natural characteristics of the recipient and its surroundings are also important. These characteristics can be classified into three different types:

1. General environmental parameters. The primary example of such a parameter is temperature. Higher temperatures increase biological activity, which can have both positive and negative environmental effects. Increased biological degradation of harmful components is generally perceived as a positive effect, while the increased BOD this causes can have negative environmental impacts.

2. Site-specific surroundings. For example, surrounding soil types affect the pH level in waterways, and hence the aquatic microbial activity.
3. Recipient-specific aspects. Water depth and turbidity are examples of such characteristics. Both are important determinants for algal growth in various water layers.

In brief, the total environmental effects on a recipient depend on multiple sets of factors and their interactions, yielding large uncertainties regarding the quality aspects of an indicator (indicated by the ellipses in Fig. 2.2).

Physical Indicators for Pollution from European Agriculture

The capacity of recipients to sustain environmental quality under pollution pressures varies throughout Europe. The most obvious difference is temperature, which rarely falls below 5°C in the southern parts of Greece, Italy, Spain and Portugal, while temperatures above this are a rare exception in the northern part of the Nordic countries for 8–9 months of the year. At the same time, nutrient loads to local recipients also vary considerably, with extreme nutrient surpluses in parts of The Netherlands and the Flanders region of Belgium (Brouwer and Hellegers, 1997). Taking a stock pollutant perspective (Conrad, 1992), it is therefore quite clear that the environmental status in local recipients differs considerably throughout Europe. In this context, an important question is which indicator(s) to choose under these varying conditions.

Several physical indicators exist for monitoring the effect of nutrients from agriculture. Each has its advantages and disadvantages. The indicators mentioned in this paper include:

1. *Input use/agronomic practices*, an extremely low-cost/low-quality indicator. While most of the necessary data are readily available, the amount of inputs used and the agronomic practices themselves may provide little information on the environmental impacts, unless coupled with extensive modelling of the nutrient leaching and erosion processes (for details on modelling such processes see Vatn *et al.*, 1996, 1997; Romstad *et al.*, 1997b).
2. *Farm-gate nutrient balances*, another low-cost indicator. In its generic setting this is also a low-quality indicator, but in settings where soils are reasonably saturated with nutrients, it may provide reasonable estimates for potential changes in environmental quality (Brouwer and Hellegers, 1997; Romstad *et al.*, 1997a).
3. *Individual farm (field) nutrient leakages*. It is generally technically difficult and costly to monitor leakages from individual farms or farm fields (Braden and Segerson, 1993). As an indicator, individual farm (field) leakages therefore implies that less information will be gathered compared to a lower-cost indicator with the same informational benefits. Its use as an indicator is therefore likely to be limited. (This is also the primary reason why recommended

policies directed at reducing nonpoint-source pollution have been oriented towards inputs and agronomic practices.)

4. *Ambient environmental quality* has traditionally been relatively costly to monitor. It is, however, one of the indicators that provides the most reliable and welfare consistent information on the environmental state of the recipient. This indicator has received little attention in a policy context as it is difficult to trace changes in ambient quality back to individual farmers.[5]

In a system where the capacity of the soil to store residual nutrients (surpluses) is not exceeded, the linkages between surpluses and emissions to the recipients depend on several factors, including soil erodibility and (water) infiltration capacity, the distribution of rainfall (surface versus intra-soil leakages), and distance to the recipient (for details, see Vatn *et al.*, 1996). Under such conditions one would not expect a close relationship between ambient environmental quality and farm-gate nutrient balances.

In a situation where the soils are nearly saturated with nutrients, and surpluses by far exceed the retention capacity of the soil for a prolonged period, changes in agricultural practices will result in reduced nutrient surpluses, but have small immediate effects on ambient quality, (4). Hence, measuring changes in ambient quality in such a setting is unlikely to provide any indication of policy effects in the short term. In terms of Fig. 2.2, this implies that the 'quality' of (4) declines, making it hard to justify its additional costs *vis-à-vis* farm-gate nutrient balances (2). At the same time, measuring farm-gate nutrient balances will provide a relevant estimate of how well the policy works with respect to changing farmer behaviour, and a reasonable and low-cost estimate of changes in nutrient loads to the recipient. As such, farm-gate nutrient balances will still not provide any consistent and reliable measure of environmental quality in the short run. In the long run, however, reduced surpluses eventually lead to reduced emissions. Measuring farm-gate nutrient balances captures this change, thereby providing parts of the information needed for predicting when emissions are reduced.

With large nutrient surpluses, little nutrient buffering capacity in the soils and limited retention of nutrients in the waterways, nutrient surpluses could produce a reliable, consistent and low-cost measure of nutrient loads to coastal waters.

Farm-gate Nutrient Balances

Under conditions where substantial nutrient surpluses have saturated the soils with nitrogen and phosphorus over a prolonged time, measuring ambient water quality is unlikely to provide any information on the effects of policy, as it may take several years for reduced nutrient leakages to result in improved water quality in the local recipients. Reduced nutrient leakages will, however, have immediate effects on coastal waters. This does not imply that keeping

track of local recipient water quality is unimportant, but that this is not sufficient. Monitoring farm-gate nutrient balances is the first step in evaluating policies aimed at improving local water quality. Once nutrient surpluses have
been substantially reduced over a long period, it is reasonable to expect local
water quality to improve, hence justifying more effort in measuring ambient
water quality frequently.

Ambient Environmental Quality

Using ambient environmental quality as an indicator raises another issue –
determination of the system boundaries, i.e. where to undertake the measurements. The argument from the previous section is that the measured effects
may vary in a local recipient (such as minor waterways) and in a larger recipient (such as the North Sea). An awareness of these differences is important in
designing the ambient quality indicator.

While farm-gate nutrient balances may provide useful information in a
system with heavy nutrient loads over a long period of time, the situation in
less-polluted regions will be quite different. Consider nutrient leakages to a
highly valuable salmon river. In such a situation, measuring farm-gate nutrient balances is unlikely to yield reliable estimates of changes in water quality.
Other factors, such as rainfall distribution and changes in infiltration capacity
(for example due to frost), together with the timing of manure or mineral fertilizer application are of higher relative importance. Sudden changes in these
variables may have large, unwanted effects on the salmon population in the
river, in particular in certain periods such as the spawning season. In this particular case, keeping track of variations of leakages throughout the year may
therefore be just as important as recording mean water quality or leakages.

CONCLUDING REMARKS

The quality of an indicator or a set of indicators includes three criteria: consistency, reliability, and predictive capacity. These are likely to vary depending
upon local conditions. By establishing a trade-off between indicator quality
and the costs associated with that indicator, a systematic and traceable criterion for selecting indicators under various situations can be obtained. An
important implication of this is that the 'optimal' indicator set is likely to differ
between locations and, as environmental quality improves, to change over
time.

By choosing an optimal or 'near optimal' set of indicators, a more favourable benefit–cost relationship in the data collection process is obtained. There
are several advantages to this, including:

- possibilities for reduced costs in the collection and processing of environmental data; and
- a potential increase in the optimal amount of information to gather, both in space and time.

An important benefit of this latter point is that it makes it easier to establish long-term targeted data series. The lack of reliable and consistent long-term environmental time-series data creates problems both for research and for the policy decision-making process.

In this chapter I have argued strongly for using differentiated and state-dependent indicators. Just as the optimal indicator set is likely to vary across countries, so are environmental standards and solutions to environmental problems. Rigid standards and command-and-control regulations are bad economics, and most likely also unfortunate from a purely environmental perspective.

NOTES

[1]This research has been triggered and funded by the Norwegian Research Council through the Centre for Soil and Environmental Research project 'Indicators for Sustainable Agriculture'. Their financial support is gratefully acknowledged, as are discussions with other project members. The standard disclaimers apply.

[2]For a review of valuation techniques see Cummings *et al.* (1986) or Mitchell and Carson (1989).

[3]Similar analyses are used in environmental economics to set optimal emission levels – the larger the marginal benefits from reduced pollution, the lower the optimal emission level.

[4]These 'quality'–cost indifference curves exhibit many of the same characteristics of mean-variance indifference curves used in EV analysis (see Machina and Rotschild, 1987, for a concise and accessible explanation).

[5]A recent attempt at coupling ambient quality with policy is described by Romstad (1997), who suggested treating farmers in watersheds as a team, making all jointly responsible. Team-orientated policies work better if farmers are relatively homogeneous and the team does not consist of too many agents. Consequently, team approaches have many practical limitations. With technological progress the cost of measuring ambient quality is likely to decrease, making ambient quality an indicator and team-orientated approaches more viable in the future.

REFERENCES

Asheim, G.B. (1994) Net national product as an indicator of sustainability. *Scandinavian Journal of Economics* 96, 257–265.

Braden, J.B. and Segerson, K. (1993) Information problems in the design of nonpoint-source pollution policy. In: Russell, C.S. and Shogren, J.F. (eds) *Theory, Modeling and Experience in the Management of Nonpoint-Source Pollution.* Kluwer Academic Publishers, Boston, pp. 1–36.

Brouwer, F.M. and Hellegers, P. (1997) Nitrogen flows at farm level across EU agriculture. In: Romstad, E., Simonsen, J.W. and Vatn, A. (eds) *Controlling Mineral Emissions in European Agriculture: Economics, Policies and the Environment.* CAB International, Wallingford, pp. 11–26.

Brekke, K.A. (1994) Net National Product as a welfare indicator. *Scandinavian Journal of Economics* 96, 241–252.

Bunn, D.W. (1984) *Applied Decision Analysis.* McGraw-Hill, New York.

Campbell, D.E. (1987) *Resource Allocation Mechanisms.* Cambridge University Press, Cambridge.

Conrad, J.M. (1992) Stopping rules and the control of stock pollutants. *Natural Resource Modeling* 6, 315–327.

Cummings, R.C., Brookshire, D.S. and Schulze, W.D. (1986) *Valuing Environmental Goods.* Rowman & Allanfield, Ottawa.

Farrow, S. and Krautkraemer, J.A. (1991) Economic indicators of resource scarcity: comment. *Journal of Environmental Economics and Management* 21, 190–194.

Hurwicz, L. (1972) On informationally decentralized systems. In: McGuire, C.B. and Radner, R. (eds) *Decisions and Organizations.* North-Holland, Amsterdam, pp. 297–336.

Machina, M.J. and Rotschild, M. (1987) Risk. In: Eatwell, J., Milgate, M. and Newman, P. (eds) *The New Palgrave: A Dictionary on Economics.* Macmillan Press, London, pp. 201–206.

Mitchell, R.C and Carson, R.T. (1989) *Using Surveys to Value Public Goods, The Contingent Valuation Method.* Resources for the Future, Washington DC.

Norgaard, R.B. (1990) Economic indicators of resource scarcity: a critical resource. *Journal of Environmental Economics and Management* 19, 19–25.

Romstad, E. (1997) *Team Approaches in Reducing Nonpoint Source Pollution.* Discussion Paper No. D-01/1997, Department of Economics and Social Sciences, Agricultural University of Norway, Ås, Norway.

Romstad, E. and Vatn, A. (1995) *Implications of Uncertainty on Eco-Eco Modelling.* Discussion Paper No. D-08/1995, Department of Economics and Social Sciences, Agricultural University of Norway, Ås, Norway.

Romstad, E., Simonsen, J.W. and Vatn, A. (1997a) Mineral emissions – an introduction. In: Romstad, E., Simonsen, J.W. and Vatn, A. (eds) *Controlling Mineral Emissions in European Agriculture: Economics, Policies and the Environment.* CAB International, Wallingford, pp. 1–9.

Romstad, E., Vatn, A., Bakken, L. and Botterweg, P. (1997b) Eco-eco modelling: the case of nitrogen. In: Romstad, E., Simonsen, J.W. and Vatn, A. (eds) *Controlling Mineral Emissions in European Agriculture: Economics, Policies and the Environment.* CAB International, Wallingford, pp. 225–248.

Vatn, A., Bakken, L., Bleken, M.A., Botterweg, P., Lundeby, H., Romstad, E., Rørstad, P.K. and Vold, A. (1996) *Policies for Reduced Nutrient Losses and Erosion from Norwegian Agriculture: Integrating Economics and Ecology.* Norwegian Journal of Agricultural Sciences, Supplement No. 23, Ås Science Park Ltd, Ås, Norway.

Vatn, A., Bakken, L., Botterweg, P., Lundeby, H., Romstad, E., Rørstad, P.K. and Vold, A. (1997) Regulating nonpoint-source pollution from agriculture, an integrated modelling approach. *European Review of Agricultural Economics* 24, 207–229.

Environmental Indicators for Agriculture: Overview in OECD Countries

3

Kevin Parris

INTRODUCTION

Since the early 1990s the Organisation for Economic Co-operation and Development (OECD) has established amongst its 29 member countries a sound foundation for work related to agriculture and environment linkages. In particular, it has:

- identified the policy-relevant concepts and issues;
- initiated an exchange of information and policy experiences among OECD countries;
- started to develop agri-environmental indicators to support policy analysis; and
- begun to examine agricultural and agri-environmental policy measures in OECD countries.

The overall objective of the OECD's work in the area of agriculture and the environment is to identify ways in which governments might design and implement policies and promote market solutions to achieve environmentally and economically sustainable agriculture at minimal resource cost to the economy and with least trade distortion. The work is undertaken against the background of growing concern about the effects of agricultural activities and policies on the environment. The achievement of sustainable agricultural production is now widely recognized by governments as a long-term policy objective in agriculture.

In most OECD countries, agriculture is a heavily assisted industry. Agricultural support policies have multiple and sometimes contradictory effects on the environment. At the same time, countries are engaged in reforms to lower

©CAB INTERNATIONAL 1999. *Environmental Indicators and Agricultural Policy* (eds F.M. Brouwer and J.R. Crabtree)

the levels of support and move towards policies that aim to provide targeted assistance to agriculture and that have the potential to be less production and trade distorting. In this context, payments to farmers for environmental purposes are increasing. Such payments are often intended to compensate farmers for the costs of reducing polluting activities or to enhance the provision of environmental services (OECD, 1998a).

In this changing policy environment, there is a need for a better understanding of the environmental effects of agricultural support, policy reform and freer trade. This concerns primarily the effects of agricultural policies on the environment, but it also applies to the impacts of environmental policies on agriculture, especially as the number of environmental measures affecting agriculture increases (OECD, 1998b).

OECD work on agriculture and the environment takes into account the commitment to agricultural policy reform made by OECD Ministers in 1987, the directions set for sustainable development at the United Nations (UN) Conference on Environment and Development in 1992, and the 1993 Agreement on Agriculture within the Uruguay Round of multilateral trade negotiations.

OVERVIEW OF RECENT AGRI-ENVIRONMENTAL INDICATOR WORK

Since the early 1990s there has been some progress in developing analytical frameworks and related indicators with the aim of monitoring the environmental effects of agriculture, and contributing to the evaluation of agricultural and environmental policies. For many OECD member countries, and non-governmental organizations (NGOs) and international agencies this work is now beginning to involve basic data collection and the measurement of indicators. Even so, there remains considerable variability in both the coverage and quality of these data and indicators across different countries, organizations and agencies.

Many OECD member countries publish periodically compendia of general environmental data, which usually encompass an agricultural segment. Significant progress has also been made in developing state of the environment reports which interpret general environmental information. Such reports have been published at least once in most OECD countries and on an annual basis in a growing number of cases. (For information on environmental monitoring and data collection in OECD countries see OECD, 1996). These reports usually include data and indicators on a number of agri-environmental related areas, such as land use, water resources, freshwater quality, wildlife habitats, landscape and soil, as illustrated for example in a recent report published by the United Kingdom (UK) Department of the Environment (1996).

Some OECD countries are developing an analytical framework in which to identify and develop indicators that address environmentally sustainable agriculture, such as Australia (Agricultural Council of Australia and New Zealand, 1993, 1996), Canada (McRae and Lombardi, 1994; McRae, 1995), Germany

(Nieberg and Isermeyer, 1994) and The Netherlands (Brouwer, 1995). Other countries have also begun to establish an analytical basis for national agri-environmental indicators (AEIs), for example France (IFEN, 1997), New Zealand (New Zealand Ministry of Agriculture and Fisheries, 1995), Switzerland (Office Fédéral de la Statistique, 1995) and the United States (US Department of Agriculture, 1994).

However, for most OECD countries the systematic collection of basic agri-environmental data and measurement of AEIs is only beginning. Moreover, where data and indicators exist, the coverage and quality vary across countries, as revealed by an unpublished OECD pilot survey of agri-environmental indicators conducted in 1995. Even so, responses to the pilot survey provide encouraging evidence of the extent and detail of basic agri-environmental data currently available and/or being developed in OECD countries from which it should be possible to calculate AEIs.

The use of AEIs as analytical tools to monitor and evaluate the impact of policies on the environment in agriculture is relatively underdeveloped. A few OECD countries have started work in this area, such as the consideration being given by Canada to develop the predictive capability of AEIs to assess the economic and environmental implications from domestic agricultural policy reform (McRae *et al.*, 1995). Eventually the Canadian analysis will also be extended to examine longer-term sustainability questions by relating anticipated changes in resource quality back to production and agricultural income projections.

A considerable number of NGOs are also engaged in agri-environmental work, both at the national and international level. At the national level, the UK NGO, Environmental Challenge Group, for example, has recently published a set of indicators to monitor the UK environment (Environmental Challenge Group, 1995), while international NGOs, such as the World Resources Institute (World Resources Institute, 1995) and the Worldwide Fund for Nature (1995) are also active in contributing to the development of environmental indicators, including indicators covering agricultural activities.

The increasing attention to environmental issues at the national level is also reflected in the work activities of various international agencies and organizations. This work is most advanced in terms of developing general environmental indicators by the UN Commission on Sustainable Development (UN, 1995), and the greater attention being paid to environmental issues in the World Bank (World Bank, 1995). At a regional level the creation by the European Union (EU) of the European Environment Agency in 1993 should also provide an impetus to the collection and analysis of environmental information across EU Member States and other European countries (see, for example, European Environment Agency, 1995).

More specific work on developing AEIs has begun in EUROSTAT in conjunction with the European Commission. EUROSTAT has organized a number of joint meetings with OECD over recent years to consider development of nutrient balances. Work is also underway in EUROSTAT on other agri-environmental areas, such as pesticide use (EUROSTAT, 1996). A project has also been recently implemented in EUROSTAT on establishing environmental indicators linked to the development of 'green' national accounts, which includes a component on agriculture (Jesinghaus, 1995).

The Food and Agricultural Organisation of the UN (FAO) has begun establishing guidelines for collecting and using indicators related to sustainable agriculture and rural development (SARD), as part of the UN Commission for Sustainable Development Agenda 21 programme to develop information systems on sustainable development (FAO, 1994). A key objective for the FAO is to develop SARD indicators to monitor and evaluate agricultural and environmental policies so as to improve policy decision-making, but mainly in developing countries.

OECD AGRI-ENVIRONMENTAL INDICATORS

Policy Context and Objectives

What is agriculture doing to the environment? And what impact do different policy measures have? These questions underlie OECD work on developing AEIs, part of a more general OECD environmental indicators effort (OECD, 1997a). The demand for information on agri-environmental linkages largely reflects the higher priority given by governments to environmental improvement. The supply of quantitative information of this sort, however, is currently inadequate. But without it governments and others cannot identify the environmental risks and benefits associated with agriculture, which makes it difficult to improve the monitoring, assessment and targeting of agricultural and environmental programmes.

The OECD set of agri-environmental indicators is intended to:

- provide information to government policy-makers and the public on the current state of the environment in agriculture, and changes to it;
- help policy-makers understand links between causes and effects and the impact of agricultural policies on the environment, and guide their responses to changes in environmental conditions;
- contribute to monitoring and evaluating policy effectiveness in promoting sustainable agriculture.

Indicators are part of a continuum ranging from basic data to indicators that usually combine data within some conceptual framework and, finally, knowledge that encompasses validated information around which a broad consensus has formed (Bonnen, 1989). The interpretation of indicators to improve our knowledge may involve simply presenting data graphically, or combining them into an index linked to some environmental target or threshold value. Indicators can also be used in the process of quantifying and/or modelling complex economic, policy and biological relationships.

The Framework for Developing Indicators and Reference Levels

The OECD work on indicators is being developed within what is called a driving force–state–response (DSR) framework (Fig. 3.1). The *driving forces* are features of agricultural practice that can cause changes in the state of the environment, such as the overuse of chemical inputs. But they may also be beneficial, such as the water-storage capacity of farming systems which can reduce problems such as soil erosion. The *state* refers to the environmental conditions that arise from these driving forces: their impact on, for example, soil, water, air and natural habitats. The *responses* refer to the reactions by farmers, consumers, the agri-food industry and government to perceived changes in the state of the environment, such as the adoption by farmers of pest-management practices that reduce pesticide use, and the use by some governments of payments to farmers to promote environmental benefits.

From a policy perspective it is crucial that a distinction can be made between those agricultural activities that benefit and those that harm the environment. This might be controversial, although whatever the reference level or baseline chosen, the direction of change of an environmental effect will indicate whether there has been an improvement or deterioration in environmental performance (Legg, 1997).

The OECD has been exploring the concept and operational use of a reference level, as an aid to the identification of appropriate policy responses (OECD, 1997b). The reference level can be thought of as the degree of social responsibility of farmers for the quality of the environment and thus it defines the distribution of property rights over environmental resources, and the level at which the 'polluter pays principle' applies. There are a number of possible ways in which the reference level could be expressed in an operational way: in terms of environmental outcomes, agricultural practices (such as codes of agricultural practice) or even in terms of levels and composition of agricultural production.

In some countries, reference levels, also referred to as target or threshold values, have been established by policy-makers to reflect the choices and standards that they wish to achieve (Adriaanse, 1993). However, comparing and assessing threshold and target values with actual values of an indicator will, in most cases, require further analysis. Deriving target and threshold values may be difficult because insufficient scientific evidence may exist for determining some environmental impacts. Also, certain target and threshold values may not be standardized within a country because of regional variation in natural endowments and environmental conditions. It should be noted, however, that the general direction of change and the range of values of the indicator over time within each country can provide useful information to policy-makers.

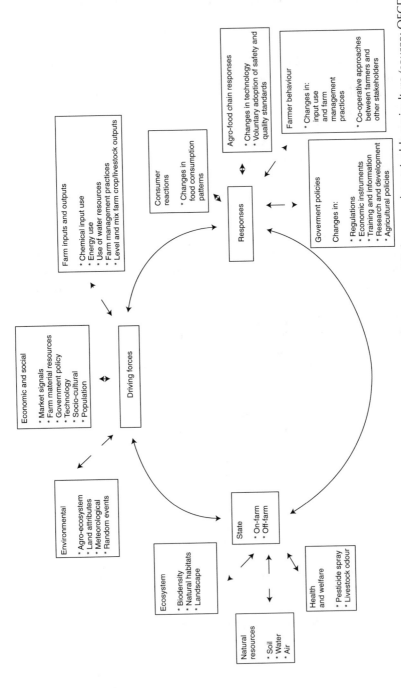

Fig. 3.1. The driving force–state–response framework to address agri-environmental linkages and sustainable agriculture (source: OECD, 1997a).

Indicators to Address Agri-environmental Issues

Indicators are being developed for the measurement of 13 agri-environmental issues, which can be grouped under three headings relating to primary agriculture (indicators might be developed at a later stage on upstream/downstream activities related to agriculture):

1. The use by primary agriculture of:

   - *nutrients* – mainly chemical fertilizers and livestock manure;
   - *pesticides* – herbicides, insecticides, fungicides and other pesticides;
   - *water* – particularly water for irrigation;
   - *land* – this covers both changing farmland use and agricultural land conservation.

2. The impact of primary agriculture on:

   - *soil quality* – the impact on soil quality, in particular to reveal the risk of erosion;
   - *water quality* – the impact on surface and groundwater quality;
   - *greenhouse gases* – both the release and accumulation of such gases;
   - *biodiversity* – of domesticated species used by agriculture, as well as of wild species;
   - *wildlife habitats* – changes and fragmentation of habitat in agricultural areas;
   - *landscape* – changes in agricultural landscapes.

3. The environmental impacts from primary agriculture related to:

   - *farm management practices* – on nutrients, pests, soil, irrigation and the farm as a whole;
   - *farm financial resources* – the varying and different sources of financial resources for farms;
   - *socio-cultural aspects* – the impact of the socio-cultural structure of rural communities.

Agricultural nutrient use

To capture how well nutrients are used in the agro-ecosystem, the OECD is developing a nutrient balance approach, which can provide an indicator of the extent to which agricultural production leads to a net surplus (or deficit) of nutrients into (or from) the soil, water or air. However, a deficit or surplus of the nutrient balance, at least over the short term, does not unambiguously indicate a beneficial or harmful environmental impact.

Several methods exist for measuring nutrient balances, all of which have limitations, partly depending on the intended level of use of the balance, ranging from the farm to the national level, and also data availability. Two main approaches are being considered to measure nutrient balances, limited at this stage of the work to nitrogen and phosphorus balances, including:

1. *The soil-surface balance*, which measures the difference between the input or application of nutrients entering the soil and the output or withdrawal of nutrients from the soil. Using nitrogen as an example, the *inputs* include chiefly the application of chemical fertilizers and livestock manure, but the use of other inputs can also be taken into account, including sewage sludge, the atmospheric deposition of nutrients in the soil (which mainly includes ammonia), the nitrogen content of crop residues remaining in fields (e.g. roots of potatoes) and the biological fixation of nitrogen by leguminous crops. The *output* consists of the withdrawal of nitrogen from harvested and fodder crops.

2. *The farm-gate balance*, which measures the difference between the nutrient content of farm inputs and the nutrient content of the outputs from the farm. Again using nitrogen as an example, the *inputs* consist of purchased materials such as chemical fertilizers, feed concentrates, manure, fodder and livestock, but natural phenomena, such as the atmospheric deposition of nitrogen in the soil and biological fixation by leguminous crops, can also be included. *Outputs* include the nitrogen content of milk, meat, manure, fodder and cereals, sold off-farm.

Preliminary OECD work on national nitrogen balance calculations, using the soil-surface method for the period 1985–1995, suggests a number of key trends in terms of the nitrogen surplus/deficit expressed as kilograms of nitrogen per hectare of total agricultural land:

- Overall the trend in nitrogen balance surpluses over the past decade is downwards or constant, especially for those OECD member countries with a nitrogen surplus in excess of 100 kg ha^{-1}.
- Generally, those countries with high livestock densities and intensive production systems have the highest nitrogen surpluses; however, regional data suggest that in areas of some countries where the national nitrogen surplus is below 100 kg ha^{-1}, they may be experiencing both the effects of nitrate pollution and soil nutrient depletion from crop production.
- Further analysis is required to explain the causes of the changing levels and trends in nitrogen surpluses/deficits between countries, and the potential environmental impacts.

Environmental and sustainable resource concerns linked to nutrient balances depend on specific agro-ecological features, hence the information derived from this indicator must be combined with knowledge about the financial and policy incentives and disincentives in place in the production system, including soil and climatic conditions, the type and density of livestock, crop production systems, farm management practices and the quality of soil and water. Thus, the nutrient balance indicator needs to be used in conjunction with information provided by other indicators, especially those covering farm nutrient management, soil quality and water quality.

In the short term, OECD work on nutrient balances would attempt to complete the current soil surface nitrogen balance calculations for all member countries. In addition, work will begin to improve the expression of the spatial variation in national nitrogen balances, particularly through the calculation of regional-level nitrogen soil-surface balances. Agreement is also being sought on: refinements to the soil surface balance methodology, such as the assumptions made on biological nitrogen fixation; how to adjust the input of nitrogen from livestock manure to account for volatilization during storage; and how crop residues left in the field after harvesting might be included in the balance.

Over the medium to long term, OECD work will explore the farm-gate nutrient balance approach. Further work would ascertain which factors, especially climatic conditions, account for changes in nutrient balance calculations, and also determine suitable reference levels from which to monitor and assess changes in nutrient use. Consideration will be given to establishing linkages with other issues, particularly soil and water quality, biodiversity and farm management practices, and also exploring a cost–benefit approach by analysing the relationship between the environmental costs of nutrient surpluses and the benefit from higher agricultural productivity.

Agricultural pesticide use
The approach being considered by OECD to measure the agricultural pesticide use issue involves classifying pesticide-use data into different environmental risk categories. This approach combines information on pesticide use with that on pesticide chemistry which influences environmental risk, that is mobility, persistence and toxicity. Until OECD evolves a suitable pesticide risk classification system, work initially would begin on collecting pesticide-use data, expressed in terms of the quantity of active ingredients per crop and/or per hectare, taking into account the proportion of agricultural land on which pesticides are applied, and the distinction between pesticide use on arable crops and pastures.

From a preliminary examination of trends in pesticide-use data in OECD countries, between 1985 and 1995, several key points emerge:

- Overall the trend in pesticide use over the past decade has remained constant or declined in most OECD countries, although pesticide use increased in a few countries.
- Variability of weather conditions may alter pesticide use: warmer conditions generally require higher uses of pesticides than colder conditions in order to maintain agricultural productivity.
- Cereals, industrial crops, fruit and vegetables account for the major share of agricultural pesticide use in most countries; while pasture and rangeland account for the major part of agricultural land use, pesticides on forage account for below 5% of total pesticide usage.

- There is considerable variation in the quantities of pesticides used per hectare both between various crops and between different countries.
- Changes in cropping systems, rotations and tillage systems can affect pesticide use, for example, agricultural land changing from arable and permanent crop use to pasture where little, if any, pesticides are used.
- Technological developments can lead to smaller quantities of active ingredients required per hectare, although the toxicity of the new products may still be high.
- The effects of organic agricultural production systems and the use of biotechnology in plant engineering in some countries on overall trends in pesticide use is at present uncertain.
- Further analysis is required to explain the causes of the changing levels and trends in pesticide use between different crops and countries and the potential environmental and health impacts.

The interpretation of pesticide-use data will be improved by linking it to an environmental and health-risk ranking system and also other indicators, in particular those covering soil and water quality and farm pest management. For example, there is some evidence that moving from more 'traditional' intensive farm practices to integrated pest management (IPM) and organic farming systems may achieve a considerable reduction in pesticide use while maintaining the economic viability of the system. On the other hand, maintaining winter green cover to limit nutrient losses from agricultural land, for example, can require the additional use of pesticides.

In the short term, OECD work will, first, improve the coverage and quality of pesticide-use data, expressed in terms of the quantity of active ingredients, particularly data on the use per crop per hectare (the OECD is currently in the process of updating and improving the quality of pesticide-use data); and secondly, develop a pesticide-use-risk ranking system that would classify pesticides used by agriculture into different risk categories according to the severity of environmental and health impacts related to exposure to pesticides. Work on the pesticide risk assessment system should be advanced by the results from an OECD Workshop on Pesticide Risk Indicators that was held in Copenhagen, Denmark, in April 1997.

In the medium to long term, OECD work might, first, establish the links with other environmental issues, especially soil and water quality, biodiversity and farm pest management; and secondly, develop a 'cost–benefit' approach by analysing environmental and health cost relationships and the benefits from pesticides through improvements in agricultural productivity.

Agricultural water use
Measurement of agricultural water use is being considered by developing water balances, both for the use of surface and groundwater resources by agriculture, together with exploring possible linkages with indicators related to farm management, especially irrigation management.

Indicators for measuring a water balance include consideration of various water use efficiency equations, monitoring stream and river flows (surface water) and also groundwater levels. Monitoring of water flows and levels might be in terms of measuring the relationship over time between surface water flows and groundwater levels in relation to rainfall within a catchment area. To monitor excess water, possible indicators could include measuring groundwater levels and the incidence of flooding. Other additional indicators being examined include calculation of water costs per tonne of crop/livestock output, and estimating the quantity of water recharged into groundwater reservoirs through certain agricultural practices.

It will be necessary to develop the water balance approach in terms of the sustainable use of water by agriculture, and also explore the linkages with indicators related to irrigation management. As regards the latter, it is necessary to identify different irrigation techniques and to classify them according to their water use efficiency in relation to a given unit of agricultural output. Work is also needed to examine the spatial aspects of agricultural water use and also consider the related issue of water resource pricing.

Agricultural land use and conservation

The indicators under consideration to monitor changes in *agricultural land use* include:

- land retired from production and maintained for conservation purposes;
- total agricultural land area in relation to the total land area;
- agricultural land per capita;
- agricultural land shifted to non-agricultural uses, including abandoned farmland;
- shifts in land use from wetlands to farmland.

Indicators under consideration to address *agricultural land conservation* cover the role of agriculture in ameliorating soil erosion, landslides and flooding, measured by the volume of:

- water stored by agricultural soils and ridges and banks (flood prevention);
- water penetrating into groundwater reservoirs relative to the outflow of water from agricultural lands into surface flows (sustainable agricultural water use);
- soil eroded from sloping land that is abandoned (prevention of soil erosion and landslides).

In further developing the indicators of agricultural land use and conservation it will be necessary to define more precisely the linkage between a particular change in land use and land conservation and environmental quality, and thus the establishment of the relevant indicators. It should be noted that understanding these linkages is also of concern to a number of other agri-environmental issues discussed in this chapter. However, this issue, and related indicators, can provide policy-relevant information on the ability of agriculture to provide environmental benefits not captured by other indicators discussed here. It will also be important to identify more clearly the links between land use and conservation and other agri-environmental issues.

Agricultural soil quality

Measurement of the impact of agriculture on soil quality is being considered through the development of a soil-risk methodology which combines indicators on the vulnerability of soil to various degradation processes, the extent of soil degradation and soil management practices.

The emphasis with this methodology is on measuring 'risk' rather than the 'state' of soil quality in view of the difficulty and cost of measuring the latter and also of separating out natural impacts from water and wind erosion from the influence of farm practices on soil quality. At present the risk method is most suited to soil erosion and salinity degradation processes, rather than aspects of soil quality such as waterlogging and toxic contamination. Estimated risk of soil degradation can be expressed in absolute terms (tonnes per hectare), classes of severity (low to excessive) or as a trend (percentage change), taking into account specific ecosystems. This provides an integration of information on the natural susceptibility of soil to change with land management practices. Although the soil-risk approach does not reveal the extent of environmental damage, it can suggest the degree of soil fragility in some regions.

While the focus is on biophysical processes of soil degradation risk, the economic consequences of degradation are also relevant. Thus, the economic effects of soil degradation could be measured through, for example, the loss of production foregone, using data for trends in yields, and the cost of the rehabilitation of soil degradation.

Further analysis is required to examine the potential of integrating indicators that address the issue of soil quality with other indicators, particularly farm management. However, some farm management practices are already incorporated into the soil-risk methodology outlined above, including the soil cover factor (a function of crops grown and the tillage practices being used). Moreover, the inclusion of some indicators that address the agricultural land use and conservation issue, such as the contribution of agriculture to the environment through its water-holding capacity, might be considered in the soil-risk methodology.

Agriculture and water quality

The method under consideration to establish agriculture's impact on water quality involves the integration of the state and risk approaches to measure surface water (rivers and lakes) and groundwater quality in agriculture:

- The state approach measures observed data on the concentrations, in weight/litre of water, of nitrogen, phosphorous, dissolved oxygen, biochemical oxygen demand, chemical oxygen demand, toxic pesticide residues, bacteria, viruses, ammonium, salinity and suspended matter resulting from agricultural activities.
- The 'risk' approach measures the ratio of potential contaminant concentration to the tolerable or allowable concentration, based on a partial budgeting method for nutrients and pesticides.

More work will be required to improve the basic data and methodologies for both the 'state' and 'risk' approaches to measure agriculture's impact on water quality. The links between this and other issues are being explored, especially nutrient and pesticide use, land use and conservation, soil quality and farm management. The need to develop methods of expressing the regional diversity concerning this issue is also necessary, including comparison against national water quality standards, and work might also examine the impact of agriculture on marine water quality drawing on other efforts in this area (GESAMP, 1990; OSPARCOM, 1994).

Agricultural greenhouse gases

To measure the release and accumulation of agricultural greenhouse gases (GHGs) the OECD is developing a net balance of the release and accumulation of carbon dioxide, methane and nitrous oxide by agriculture, expressed in CO_2 equivalents. The net balance method of measurement can provide a better reflection of agriculture's contribution to climate change than just measuring gross emissions (mainly methane and nitrous oxide), by taking into account the role of agricultural GHG sinks, in particular agricultural soils. The need to ensure consistency with other international methodologies to calculate GHG in this area is important. The net balance GHG measurement will require refinement in terms of including the use of fossil fuel on farms, and clarification of the distinction between farm-forestry and forestry. It will also be necessary to provide information that can reveal the range of uncertainty in estimates of agriculture as a sink and source for GHG, shown by some studies to be substantial. Consideration also needs to be given to examining links with the farm management indicator to measure the options to reduce GHG emissions and/or develop sinks in agriculture.

Agriculture and biodiversity

The development of indicators to address biodiversity in agriculture is complex because of differing levels at which it operates in agriculture. Since it is possible to preserve biodiversity *ex situ* and *in situ*, the indicators that could address biodiversity in agriculture will need to reflect both approaches, including the measurement of the biodiversity of 'domesticated' species in agriculture and the impact of agriculture on the biodiversity of 'wild' species.

It will be necessary to define more clearly the link between agriculture and biodiversity, in particular drawing a sharper distinction between biodiversity of 'domesticated' species and the biodiversity of 'wild' species. Work will also need to focus on the significance of site specificity, as the scale of many biodiversity issues is at the sub-national level. Methods also need to be evolved that can interpret the environmental impact of agriculture on biodiversity, such as whether a farm structure with smaller field plots and a denser and more comprehensive network of border elements, such as hedges and boundary strips, provides more favourable conditions to enhance biodiversity. Work will also explore links between this issue and those covering agriculture in relation to wildlife habitats and landscape.

Agriculture and wildlife habitats

Indicators to measure agriculture and wildlife habitat are not yet established, but the following are under consideration: changes in the area of selected 'large-scale' habitats in agriculture, such as wetlands and pasture; fragmentation of habitats both in the agro-ecosystem and natural habitats; length of 'contact zone' between agricultural and non-agricultural lands.

None of these indicators provides a direct causal link between agricultural activities and impacts on habitats, although if used in conjunction with other indicators, such as those that address nutrient and pesticide use and farm management, they may contribute information to establish these linkages. The indicators outlined here provide little information on the relationship between changes in the quality of habitats and agriculture, although the measurement of changes in 'key indicator' wildlife species and habitat fragmentation could be of value in this context.

In developing indicators to address the link between agriculture and wildlife habitats, work is required to define more precisely the scope of wildlife habitats in agriculture and establish the linkages between natural habitats and agriculture, especially in relation to key indicator species and landscape. National legislation and international agreements may be of help in establishing these definitions and linkages. More work has to be undertaken to understand the direction of agriculture's impact on biodiversity. The difficulty of encapsulating the spatial diversity of wildlife habitats into national wildlife habitat indicators also needs to be addressed.

Agricultural landscape

Indicators to measure the complexity and diversity of agricultural landscapes have not yet been established by OECD, but some member countries are beginning to develop indicators, including the measurement of landscape through:

1. *Estimating the monetary value of landscape,* using economic non-market valuation techniques, such as the contingent valuation method. There remain many conceptual and practical difficulties with these techniques as they can be resource intensive when applied on a large scale, and involve subjective judgements.

2. *Developing an inventory of physical landscape features*, such as the linear distance of hedgerows, monitoring trends in land use and appearance of key species. A difficulty with this approach is the choice of features or key species to include in the inventory (which will vary among countries and regions within countries), and the problem of assessing whether a change in the inventory represents a positive or negative environmental impact, related to agricultural activities.

Further work will be necessary to understand the linkages between landscape, agriculture and the environment, and to determine appropriate indicators. Additional work might also examine the links between landscape and those covering agriculture and biodiversity and wildlife habitats, by analysing these links at different levels, including species and genes at the lower level, habitats and man-made landscape features at the middle level, and landscape at the upper level.

Farm management

A number of indicators to assess the environmental impacts of farm management practices are under consideration, including the measurement of:

- *nutrient management* – the share of land which is analysed regularly for soil phosphorus; the share of farms using a nutrient management plan; the areas of land which require less than the normally recommended nutrient inputs and also receiving excessive (i.e. above recommended levels) of nutrient inputs; the timing of slurry application and months of available slurry storage on farm; and the use of low ammonia emission slurry application machinery;
- *pest management* – the share of land on which IPM practices are adopted; the use of pest forecasting systems; the area of cropping land on which pesticides are not applied; and the efficiency of pesticide spraying equipment in applying pesticides;
- *soil management* – the share of land on which soil conservation practices are adopted, including the use of winter cover crops and appropriate tillage practices;
- *irrigation management* – the efficiency of water use on irrigated land in terms of the quantity of water used to produce a unit of agricultural output; and the pricing of water to agriculture;
- *whole farm management* – the rate of adoption of farm plans or property management plans – which, when fully developed, may contain information relating to economic, farm production and biophysical or environmental factors – either approved by governments or voluntarily.

Further analysis of farm management practices will be required in terms of defining measurable indicators that can separate out environmentally appropriate from inappropriate practices. Additional work could also explore the relationship with other agri-environmental issues and related indicators. As some countries have already defined certain practices in national legislation, for example 'organic farming', while the FAO has developed internationally accepted definitions for certain farm management practices, such as IPM, this information could provide an input to develop indicators that address the farm management issue.

Farm financial resources
The indicators under consideration to address the issue of farm financial resources and the environment include measurement of: net farm and off-farm income; policy transfers; average rate of return on capital employed; and the average debt/equity ratio, on a per farm basis and adjusted for inflation in real terms. More work has to be completed on defining the direction of environmental impact associated with changes in the level of farm financial resources. Further investigation is also required of the links between farm financial resources, farm management practices employed and the effect on the environment, taking into account other factors, such as longer-term climatic changes and population growth, which may indirectly influence farmer behaviour and environmental outcomes.

Socio-cultural issues in relation to agriculture
Although the importance of socio-cultural issues in the analysis of agriculture and the environment, including sustainable agriculture, is generally accepted, no precise definition of the policy issues nor relevant indicators have yet been established. However, some indicators are under consideration, including the measurement of:

• land-use changes, especially the transfer of agricultural land to use for urban development;
• changes in population growth and composition, in particular rural–urban changes;
• education and training of farmers, in relation to the adoption of environmental plans and sustainable farming practices;
• farmer health and safety, related to the use of agricultural pesticides and machinery.

The key aspect to further work to address the socio-cultural issue in relation to agriculture and the environment relates to establishing a clear definition of the relevant policy issues and developing indicators that can quantify these issues. Work under way both on other agri-environmental issues, for example land use and farm management, and in other OECD activities, such as on rural development and structural adjustment in agriculture, could be drawn on to help develop the definition of issues and identification of indicators. However, further conceptual work is also required to examine the linkages between the socio-cultural, economic and environmental components in the DSR framework described in this chapter (Fig. 3.1), and also to consider the spatial aspects of these links in developing appropriate indicators.

FUTURE OECD AGRI-ENVIRONMENTAL INDICATOR WORK

Building on recent progress to date, the OECD agri-environmental indicator work is expected to advance analysis, over the short to medium term, in the following areas:

1. Improvement of the conceptual and analytical understanding of the links between agriculture and the environment in specific areas, to help identify which policy-relevant indicators might be developed and how they should be measured; in particular, for example, biodiversity, habitats, landscape, farm financial resources and socio-cultural aspects of agri-environmental linkages. In these areas the links between agriculture and environmental impacts need further refinement and the identification of relevant indicators.
2. Identification of policy-relevant indicators and methods of measurement for those agri-environmental issue areas where the conceptual basis is advanced but for which indicators and methods of measurement are not yet established, such as pesticide use, soil and water quality.
3. Collection, systematically, of basic agri-environmental data and beginning the calculation of indicators where methods of measurement are established. In this regard, work is already under way on calculating indicators and collecting basic data related to, for example, nutrient use and GHGs.
4. Examination of how the DSR framework and related indicators can be used as analytical tools to better understand agri-environmental relationships in policy analysis and to evaluate the impact of policies on the environment in agriculture.

The OECD has developed a framework in which to analyse agri-environmental linkages, using the 'driving force–state–response' model (Fig. 3.1). However, the quantification of these relationships is at too early a stage to draw firm conclusions on agriculture's environmental performance across all of the key areas. Many of the environmental effects can take a long time to become apparent. Some of the trends may move in a 'positive' direction at the same time as others move in the opposite direction. The indicators are 'physical' measures, and interpreting changes in a range of different measures is difficult. A fuller picture will result when other indicators, such as those for water quality, wildlife habitats and landscape, become available.

The indicators being developed promise to provide valuable information by revealing where a problem may be emerging that might require a policy response, and as a contribution to monitoring the effects of actions taken by farmers in response to changing policy incentives or disincentives. This work could also throw some light on quantifying the causes of changes in environmental trends, which could contribute to understanding possible future environmental developments under different scenarios. Changes in technology, the financial situation of farmers, socio-cultural influences, agricultural and agri-environmental policy measures, and environmental regulations, all combine to influence the level, composition, location and practices of farmers that determine the environmental outcomes. Indicators for farm management practices, for example, should help in understanding these linkages, although matching them to environmental outcomes is a complex task.

There is a need to explore the possibilities of incorporating the physical data into an economic accounting framework in order to further explore many of the policy questions and linkages. For example, the costs incurred by farmers, or other economic sectors, or by public bodies in tackling pollution from agricultural activities, could be combined with physical indicator data. Or the costs incurred in preserving landscape features could be weighed up against other evidence of the public's valuation of landscape benefits. However, the objective valuation of non-marketed attributes, such as the preservation of landscape, is more difficult than measuring the value of water or soil quality, for example. The relationships between the environmental costs from agricultural activities, the benefits from agricultural production and the level of agricultural support need further analysis.

NOTE

The views expressed do not necessarily reflect those of the OECD or its member countries.

REFERENCES

Adriaanse, A. (1993) *Environmental Policy Performance Indicators – a Study on the Development of Indicators for Environmental Policy in the Netherlands*. Ministry of Housing, Physical Planning and the Environment, The Hague.

Agricultural Council of Australia and New Zealand (1993) *Sustainable Agriculture: Tracking the Indicators for Australia and New Zealand*. Report prepared for the Standing Committee on Agriculture and Resource Management, CSIRO Publications, Victoria, Australia.

Agricultural Council of Australia and New Zealand (1996) *Indicators for Sustainable Agriculture: Evaluation of Pilot Testing*. Report prepared for the Sustainable Land and Water Resources Management Committee, Standing Committee on Agriculture and Resource Management, CSIRO Publications, Victoria, Australia.

Bonnen, J.T. (1989) On the role of data and measurement in agricultural economics research. *Journal of Agricultural Economics Research* 41 (4), 2–5.

Brouwer, F.M. (1995) *Indicators to Monitor Agri-environmental Policy in the Netherlands*. Agricultural Economics Research Institute, The Hague.

Department of the Environment, UK (1996) *Indicators of Sustainable Development for the United Kingdom.* DoE, London.

Environmental Challenge Group (1995) *Environmental Measures: Indicators for the UK Environment.* Environmental Challenge Group, UK.

European Environment Agency (1995) *Europe's Environment: The Dobris Assessment.* Office for Official Publications of the European Communities, Luxembourg.

EUROSTAT (1996) *Overview of Pesticide Data in the European Union.* Statistics in Focus – Environment, No.1. EUROSTAT, Luxembourg.

FAO (1994) *Agricultural Policy Analysis and Planning: the Use of Indicators to Assess Sustainability.* Paper presented to the OECD Meeting of Experts on Agri-environmental Indicators, Paris, 8–9 December.

GESAMP (1990) *The State of the Marine Environment.* GESAMP – Joint Group of Experts in the Scientific Aspects of Marine Pollution (IMO, FAO, UNESCO, WHO, IAEA, UN, UNEP). Blackwell Scientific Publications, Oxford.

IFEN (1997) *Agriculture et environnement: les indicateurs.* Institute Français de l'Environnement, édition 1997–1998, Orléans.

Jesinghaus, J. (1995) *Green Accounting and Environmental Indicators: the Pressures Indices Project.* Paper presented to the SCOPE Workshop on Indicators of Sustainable Development, Wuppertal, Germany, November.

Legg, W. (1997) *Have OECD Agricultural Policy Reforms Benefited the Environment?* Paper presented at the UK Agricultural Economics Society Annual Conference, Edinburgh, 21–24 March.

McRae, T. (1995) *Report of the Second National Consultation Workshop on Agri-environmental Indicators for Canadian Agriculture.* Agriculture and Agri-Food Canada, Ottawa.

McRae, T., Hilary, N., MacGregor, R.J. and Smith, C.A.S. (1995) *Design and Development of Environmental Indicators with Reference to Canadian Agriculture.* Report No. 9, Environment Bureau, Policy Branch, Agriculture and Agri-food Canada, Ottawa.

McRae, T. and Lombardi, N. (1994) *Report of the Consultation Workshop on Environmental Indicators for Canadian Agriculture.* Environment Bureau, Policy Branch, Agriculture and Agri-Food Canada, Ottawa.

New Zealand Ministry of Agriculture and Fisheries (1995) *Proceedings of the Indicators for Sustainable Agriculture Seminar.* MAF Policy Technical Paper 95/7, August. Wellington, New Zealand.

Nieberg, H. and Isermeyer, F. (1994) *The Use of Agri-environmental Indicators in Agricultural Policy.* Paper presented to the OECD Meeting of Experts on Agri-environmental Indicators, Paris, 8–9 December.

OECD (1996) *Integrating Environment and the Economy – Progress in the 1990s.* OECD, Paris.

OECD (1997a) *Environmental Indicators for Agriculture.* OECD, Paris.

OECD (1997b) *Environmental Benefits from Agriculture – Issues and Policies. The Helsinki Seminar.* OECD, Paris.

OECD (1998a) *Agriculture and the Environment: Issues and Policies.* OECD, Paris.

OECD (1998b) *The Environmental Effects of Reforming Agricultural Policies.* OECD, Paris.

Office Fédéral de la Statistique (1995) Les plantes, les animaux et leurs habitats. *Statistique suisse de l'environnement,* No. 2.

OSPARCOM (1994) *OSPARCOM Guidelines for Calculating Mineral Balances.* Working Group on Nutrients, Oslo and Paris Conventions for the Prevention of Marine Pollution (OSPARCOM), NUT 94/8/1-E, Berne, Switzerland.

UN (1995) *Information for Decision-Making and Earthwatch.* Economic and Social Council, Commission on Sustainable Development, E/CN.17/1995/7 February. UN, New York. Information for decision-making. In: *Indicators of Sustainable Development Framework and Methodologies.* UN, New York, pp. 411–418.

US Department of Agriculture (1994) *Agricultural Resources and Environmental Indicators.* Economic Research Service, Natural Resources and Environment Division, Agricultural Handbook, No. 705. US Department of Agriculture, Washington DC.

World Bank (1995) *Monitoring Environmental Progress: A Report on Work in Progress.* Environment Department, Washington DC.

World Resources Institute (1995) *Environmental Indicators: a Systematic Approach to Measuring and Reporting on Environmental Policy Performance in the Context of Sustainable Development.* World Resources Institute, Washington DC.

Worldwide Fund for Nature (1995) *Measuring Progress Toward Bio-Intensive IPM: a Methodology to Track Pesticide Use, Risks and Reliance.* WWF, Gland, Switzerland.

Agricultural Sector Pressure Indicators in the European Union

4

Jochen Jesinghaus

INTRODUCTION

Following discussions among environmental experts of various General Directorates of the European Commission, which resulted in the Communication from the Commission to the Council and the European Parliament on *Directions for the EU on Environmental Indicators and Green National Accounting* (COM (94) 670 final, 21.12.94), a series of pressure-indicator projects was launched in 1995. These activities served to establish a harmonized European System of Environmental Pressure Indices (ESEPI), aiming at a detailed physical description of harmful human activities at the aggregation level of 'policy fields'[1] such as *climate change*, *loss of biodiversity* or *dispersion of toxic substances*.

This set of ten indices will describe pressures in a highly aggregated format. The goal of aggregating pressure indicators to indices is to 'build the bridge' between environmental scientists and decision-makers. Pressure indices are designed to communicate environmental information to a not-so-specialized audience – that is, the general public and decision-makers in non-environmental policy – without losing the scientific soundness of the original indicators.

Within the logic of the pressure–state–response model, links to *state* and *response* indicators will be given particular attention. A flexible framework will allow the system to be adapted to the latest state of the art. Compatibility with approaches to extend the System of National Accounts (SNA) with 'Green' Satellite Accounts will be sought in so far as the consistency of the pressure index is not compromised from the viewpoint of environmental sciences. The final goal is to produce a tool that serves environmental policy in very much the same way as the SNA has served economic policy in the second half of this century.

In autumn 1997, work on an 'Early Harvest Publication' began with the goal of producing the first set of 60 pressure indicators covering the European Union (EU) in 1998.

DEFINITION OF USER DEMAND THROUGH EXPERT SURVEYS

From the very beginning, it was clear that in spite of the indicator work already done, e.g. in the Organisation for Economic Co-operation and Development (OECD) indicator groups (see Chapter 3), the ambitious goal of producing a comprehensive picture of pressures on the environment conflicted with the quality and coverage of the necessary database. Our own optimistic estimate was that we could derive approximately 40 pressure indicators from data that are available on a European scale. However, a satisfactory coverage of the policy field 'environment' would require about 100 indicators and, further- more, many of the 'environmental' indicators, such as *pesticide use* (expressed in tonnes per year) or *total waste production*, are mere adaptations of economic statistics that would not find the approval of environmental experts. Last, but not least, most of the available data and indicators are, at least in comparison to economic indicators such as gross national product or the unemployment rate, so out of date as to be of more interest to historians than to decision-makers.

Improving the database will require heavy investment not only in meth- odological studies,[2] but also in data collection. Partly, this task could be done by the European Environment Agency (EEA), especially in those areas where the expertise of natural scientists is required, e.g. in the development of state indicators and appropriate models to link state to pressures. However, most data requirements for *pressure* indicators correspond better to the profile of Member States' statistical services, since the linkage to standard socio- economic statistics is essential in order to provide up-to-date indicators with a sectoral breakdown (as requested by the Fifth Environmental Action Prog- ramme). In order to ensure that the burden on our partners is minimized, and that the data collection efforts are justified by user demand, EUROSTAT has explored users' views through two expert surveys conducted in December 1995 and October 1996.

The first questionnaire was sent out in December 1995 to over 2300 Euro- pean environment experts, of which over 600 replied. The expert panels, called Scientific Advisory Groups (SAGs), had been selected by ten specialized insti- tutes (SIs) for the policy fields:

- *climate change*;
- *ozone layer depletion*;
- *loss of biodiversity*;

- *resource depletion*;
- *dispersion of toxic substances*;
- *waste*;
- *air pollution*;
- *marine environment and coastal zones*;
- *water pollution and water resources*;
- *urban environmental problems*.

The main criteria for the selection were: (natural) scientist with a long professional experience in the respective field; equal coverage of all 15 EU Member States and political sectors. The institutes used public sources such as workshop participant lists and national research programmes to find these experts.

The first questionnaire was deliberately simple: the introduction describing the OECD's pressure–state–response model was followed by a blank page on which the expert was asked to list five important pressure indicators for their policy field. The intention was to avoid giving the panel a 'pre-defined' set, but rather to invite them to be imaginative, and to look as far as possible into the future. Statistical work takes time, and we wanted to receive an 'early warning' from the scientists on which would be the 'hot issues' on the political agenda at the time when our efforts would come to fruition.

One of the lessons learned from the first survey was that many indicators are needed. The respondents each listed about 4–5 indicators, and thus a total of 2700 indicators were suggested. Even after excluding duplication, the specialized institutes in charge of each policy field still had to deal with more than 1000 extremely wide-ranging indicator proposals. This huge and heterogeneous set of suggested indicators was reduced to a manageable set of 10×30 indicators, which was used as the basis for the second survey in October 1996. This survey aimed at obtaining a differentiated assessment of the list of 300, in order to facilitate the mid-term choice of a European core set of around 60–100 pressure indicators.

The following questions (abbreviated) were posed for the panels:

1. Policy relevance. How important would the pressure indicators presented here be to a national policy-maker, e.g. in the environment ministry of your country? *[ranking 'very low' = 1, 'very high' = 4]*

2. Analytical soundness. How strongly do you consider changes in each *indicator* presented here would correlate with *real* changes in the environmental pressure affecting this policy field?

3. Response elasticity. Given the present technical and economic obstacles, how difficult would it be to take action to significantly reduce each pressure *indicator* (e.g. by 5%)? Or, in other words, how easily ('elastic') could decision-makers respond to a particular pressure indicator showing a negative trend?

4. Core indicators. Imagine that you would have to describe the overall pressure in this policy field using a maximum of *five* absolutely essential indicators from the list presented here; which five would you choose? Ideally, your 'core set' should be based on your evaluation in Questions 1–3.

The results of the second survey provided a transparent picture of information needs. The 'quality' questions (1–3) allowed an assessment to be made on where additional work is needed to improve the definition and usefulness of these indicators. The 'core indicators' question, on the other hand, forced the experts to 'sacrifice the good for the essential' – which reflects the situation of data providers with limited resources who have to make a choice. The selection of the first 60 pilot indicators has indeed been based primarily on the results of Question 4.

In the following section, some of the results are presented and pressure indicators of specific relevance for the economic sector agriculture are discussed. Figure 4.1 shows the results for the 15 top-scoring indicators of the policy field, *loss of biodiversity*. It is noteworthy that an 'exotic' indicator such as *wetland loss through drainage* scores second in the core list, chosen by more than 50% of the biodiversity experts.

In contrast to the 'flat' ranking above, the policy field *water pollution and water resources* is dominated by a few indicators: nutrient use, groundwater abstraction and pesticide use were each chosen by more than half of the water experts (Fig. 4.2 right-hand scale). It is noteworthy that the experts differentiated between 'policy relevance' (by giving rank 1 in *water pollution and water resources* to pesticides) and the 'core indicators' question (where nutrients

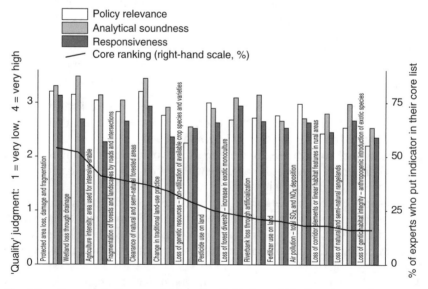

Fig. 4.1. Pressure indicators of the policy field, *loss of biodiversity*: quality questions Q1–3, core ranked.

scored highest; *all* experts included them in their list). While they consider nutrients overall more important, they also recognized the prominent role of pesticides in water policy.

The other indicators in the 'water top ten' are less relevant for agriculture (for more details see Bittermann and Brouwer, 1998). The situation in the other policy fields is as follows:

- **Air pollution**: *ammonia emissions* rank 7, *pesticides*[3] rank 9.
- **Climate change**: *methane emissions* score second after *carbon dioxide emissions*, followed by *nitrous oxide emissions*. The sector agriculture, usually associated more with water or biodiversity problems, has thus been given a surprisingly prominent position in *climate change*.
- **Marine environment and coastal zones**: *eutrophication* scores highest, *priority habitat loss* follows on rank 4, and *wetland loss* occupies rank 8 in the list. However, the latter pressures are not necessarily attributable to agriculture – the experts put *tourism intensity* in the list, thus explicitly blaming another economic sector for the environmental problems in this area.
- **Ozone layer depletion**: the pesticide *methyl bromide* appears on rank 7; *methane* and *nitrous oxide*[4] follow on ranks 8 and 9.
- **Resource depletion**: *water consumption per capita* scores highest. The *nutrient balance of the soil*, expressed as nutrient input/nutrient output in %, appears on rank 4; *groundwater abstraction for agricultural/industrial*

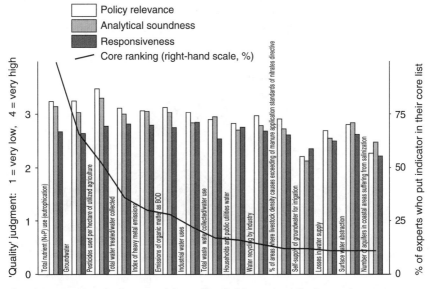

Fig. 4.2. Pressure indicators of the policy field, *water pollution and water resources*: *quality* questions Q1–3, core-ranked.

..poses on rank 10 – in some arid regions of the EU, agriculture may be the main groundwater user.

- **Dispersion of toxic substances**: _consumption of pesticides by agriculture_ is the indicator with the highest score.

Finally, the two policy fields _urban environmental problems_ and _waste_ have no agriculture-specific indicators in their top ten list.

DECISION-ORIENTATED INDICATORS: THE EXAMPLE OF _PESTICIDE USE_

In the 'water' field, _pesticide use_ ranked third, but it was attributed the highest overall policy relevance. The same indicator appears in the policy fields _loss of biodiversity_ (rank 8), _air pollution_ (rank 9), and, last but not least, in _dispersion of toxic substances_ (rank 1). In the original water questionnaire, the indicator was defined as: _pesticides used per hectare of utilized agriculture area, measured as toxicity equivalents/ha_. This may explain why the water experts also attributed a high analytical soundness to this indicator – our 'real' figures still express pesticide use in simple tonnes. Everybody will agree that this is analytically rather unsound – the environmental risks are only loosely correlated with the total amounts applied. But what are the real risks? In the following, I will try to portray the decision-tree associated with political decision-making on pesticides (hoping that pesticide experts from both the farmers/pesticide producers[5] and the non-governmental organization (NGO) side of the political spectrum can locate themselves somewhere in the scheme).

The highest level of the discussion of environmental problems (Fig. 4.3) is a political controversy between the 'societal actors', e.g. industry and farmer associations, environmental NGOs and the government. Triggered by sensational stories in the media ('we are being poisoned day by day'), these actors

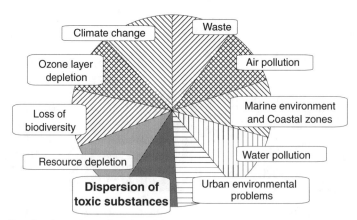

Fig. 4.3. Top level: environmental problems.

might turn their attention to one particular policy field, e.g. dispersion of toxic substances (Fig. 4.4). At this level, discussions tend to ignore details and science.

When discussion at the political level has ended with the conclusion 'We must stop poisoning our voters', it is handed over to technical committees. Toxicologists from both sides of the political spectrum try to find out why the toxic substances are on the agenda again, and they may try to assess soberly what the real dangers are. Is it a question of dioxins from steel production or is the daily use of household chemicals more dangerous? The pesticide issue will not be neglected, but this specific problem will be handed over to a panel of pesticide experts.

They will have a closer look at the 'big three' in this cluster of toxic chemicals. At this stage, scientific considerations become more relevant. For example, the reader may ask why the insecticides have been given such a big share in Fig. 4.5. The answer is simple: the graph is purely illustrative, and the shares have been chosen arbitrarily; for a pesticide index based on toxicity considerations, the shares of the pie chart would have to be calculated with the

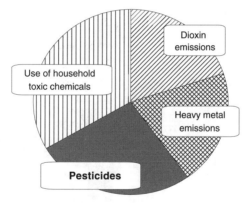

Fig. 4.4.   Level 2: toxic substances.

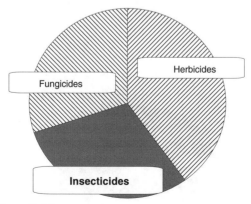

Fig. 4.5.   Level 3: pesticides.

help of toxicity equivalents, etc. But even that level of detail would not be sufficient.

To describe the impact of insecticide use properly, more detail is needed (see Fig. 4.6). For example, it matters where the insecticides are being applied. Will they kill agricultural pests only, or will the disappearance of the insects shatter the balance of a sensitive ecosystem nearby? Is the groundwater level so high that leaching can be expected? How do the rainfall patterns influence the leaching to groundwater? And, first of all, doesn't the danger very much depend on how cautiously the insecticides are being applied?

Toxicity, persistence and other 'scientific' criteria only refer to a small part of the problem. It may be more relevant whether pesticide residues remaining in the spraying equipment end up dispersed over the field, or in a nearby watercourse. It matters how well trained the farmers are, and whether they apply pesticides strictly according to the rules – and instruction manuals play an important role in helping to ensure that this occurs. It matters whether pesticides are so cheap that they are used 'preventatively', or expensive enough that they are being used only in cases of emergency (see Fig. 4.7).

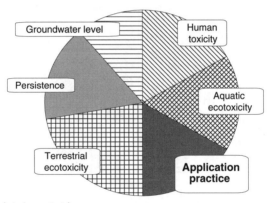

Fig. 4.6.   Level 4: insecticides.

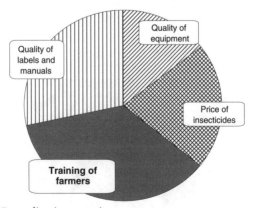

Fig. 4.7.   Level 5: application practice.

Constructing a pesticide index that properly reflects environmental pressures caused by the use of pesticides would have to follow this decision tree closely. Many of the data needed, however, are not available as a statistical series: they probably exist somewhere in the drawers of government institutions, since registering a pesticide is a long process during which all the considerations described above play a role. The 'mental' model of pesticide use and the associated risks and impacts exists in the heads of those government officials and private experts who are involved in pesticide registration. Translating their mental model into a mathematical model that condenses all contributions to 'the pesticide problem' into pesticide risk indices[6] is not a trivial task, since many of the data necessary are not available. EUROSTAT, together with national statistical institutes, have begun a number of projects[7] aimed at supplying and improving the basic data required for a more accurate description of pesticide use problems, for example:

- a database linking trade names to active ingredients;
- the development of guidelines for the collection of data on pesticide use; and
- the improvement of existing apparent consumption/sales data.

Obviously, the data situation[8] does not yet allow production of a perfect pesticide risk index that would give guidance to policy-makers through all levels of the decision-tree. Some of the data requirements for an ideal index concern state indicators and area value indices[9] which still have to be defined and developed.

DISCUSSION

One of the conceptually most interesting questions is perhaps whether the best should be the enemy of the good. Should we wait until all the detail is available, and in the meantime continue to use tonnes as a measure of pesticide risk? Or should we try to move down along the decision-tree, adding level by level a more detailed description?

A good indicator should, of course, tell the truth (but unfortunately, there are many scientific truths). In a sense, tonnes are true – they are measurable, and the figure itself is accurate in a statistical sense. Beyond 'true' figures, however, one should look at the way indicators *function*. Any published indicator exerts a pressure on decision-makers to improve the figure – but not necessarily to improve the underlying *problem*: a pesticide use indicator expressed in tonnes, for example, forces decision-makers to reduce the amounts of pesticides used irrespective of their toxicity, and of all the other factors that make pesticide use an environmental problem. Tonnes are misleading, they push the decision-makers in the wrong direction.

This *true figure, wrong message* problem is common to most indicators that aggregate heterogeneous substances on the basis of mass units. One could portray fertilizer use in very much the same way as the example above – the effects of fertilizers depend on so many circumstances that aggregation to a simple tonnes per hectare figure would be misleading.

Producing decision-relevant indicators is a difficult and costly exercise. Agriculture experts have gone a long way, e.g. with the development of harmonized nutrient balance models. Pesticide risk indices will certainly follow, given the high demand for them.

The main value of the expert surveys conducted by Eurostat is perhaps that the identification of the most important 50–100 indicators through the ranking process can help us to direct our efforts towards those which are not yet on the agenda of the 'indicator community', but whose absence will be felt when environmental policy incorporates the services that the newly developed 'advanced' indicators provide. One may think that producing a hundred good environmental indicators is 'expensive'. One should first compare the environmental data situation to the detail that is provided, for example, in production statistics. A second argument might be more convincing: the rich societies of the EU are thought to spend about 1–2% of their gross national product (GNP) on environmental protection. It is almost impossible to give an accurate figure in monetary terms for the damage done by the remaining pressures, but protection expenditures and environmental damages together may well exceed 5% of GNP. Given the lack of rationality of environmental policy, it should be easy to reduce this figure drastically through indicators that make the complexity of this policy field more transparent. Better indicators and indices can help the opposing parties to end their 'religious wars', and to concentrate on the question: 'how can we achieve a maximum of environmental protection with a minimum of costs to the economic sectors?'

NOTES

[1] The ten policy fields have been chosen to correspond closely to the seven 'themes' of the Fifth Environmental Action Programme (93/C 138/01, *Official Journal of the EC*, 17.5.93, p. 42ff.).

[2] The pressure indices project was started with more than 30 sub-projects, performing three basic functions:

- Ten pilot projects (following the ten policy fields) are aimed at identifying the 'demand' for indicators and related statistics from the standpoint of the users (working with expert panels as described above).
- Twelve sectoral infrastructure projects (SIP) were created to build up the methodological 'infrastructure' for the calculation of pressure indicators. These projects follow the structure of the 'target sectors' of the Fifth Environmental Action Programme – Towards Sustainability (energy, agriculture, transport, industry, tourism and waste management)
- Currently eight 'environmental pressure information system' projects (EPIS), aimed at 'supplying' indicators through a European database, which integrates

environmentally relevant physical and economic data, and facilitating the access of indicator experts to such data.

[3] Some of the results, e.g. the appearance of pesticides in 'air pollution', may result from the fact that the SAG panels acted independently of each other, and thus included indicators of general importance but of little relevance for their specific field.

[4] See previous footnote. At first sight, the greenhouse gases methane and N_2O in the policy field 'ozone layer depletion' seem to be another case of 'caring for another policy field'. There is some evidence, however, that greenhouse gas-induced temperature rise also promotes ozone depletion through the resulting cooling of the lower stratosphere. In addition, methane and N_2O react chemically with other ozone-relevant substances.

[5] These reflections are inspired by a discussion with Ulrich Ramseier, Novartis Crop Protection AG, Switzerland.

[6] Ideally, there should be separate indices for the policy fields 'dispersion of toxic substances' (with human health as the central concern); 'water pollution and water resources'; and 'loss of biodiversity' (the latter being more concerned with the effects of pesticide use on the 'health' of ecosystems).

[7] Contact: Rosemary Montgomery, EUROSTAT/F3.

[8] Often, data are available in principle, but either difficult to access or not comparable enough for use in indicators. This is particularly the case for the biodiversity indicators. The EEA has offered to EUROSTAT the assistance of its topic centre on nature conservation (ETC/NC, Paris) in further developing 'loss of biodiversity' indicators and providing data for their calculation.

[9] Many environmental pressures only become *problems* if the area affected is of a particular value, e.g. as a groundwater reserve, for the beauty of its landscape, its use as a recreation area, or because of its richness in rare species. Theoretically, this value could be expressed in monetary terms, based on willingness-to-pay, etc. However, many scientists will feel that a description of the physical aspects (e.g. with state indicators) that were considered in the valuation, aggregated in a transparent way to non-monetary value indices, would be more helpful than 'putting price tags on butterflies'. Such indices could then be used to 'qualify' pressures by weighting them against the value of the affected area. As long as such indices do not exist, differentiation by normal and protected areas would be a first proxy.

REFERENCE

Bittermann, W. and Brouwer, F. (1998) *Joint Final Report of the Sectoral Infrastructure Project* (SIP), Agriculture, 1998.

Agri-environmental Indicators in the European Union: Policy Requirements and Data Availability

5

Floor Brouwer

INTRODUCTION

Agri-environmental indicators are increasingly required to monitor progress on the achievement of environmentally friendly production methods by the agricultural sector. Targets are formulated in the context of environmental legislation and agricultural policies, and aim to contribute to the adoption of less intensive production methods. The objective of such indicators is essentially twofold (Brouwer, 1995).

The first is to assess effects on the environment, nature and landscape in response to changes in the agricultural sector. Such indicators, however, would require monitoring of (i) use of chemical inputs in the agricultural sector; (ii) emission levels related to farming; and (iii) landscape and nature implications of agricultural practice. They might also explore the extent to which the agricultural sector meets environmental targets and contributes to beneficial impacts on the environment, landscape and nature. Essentially, such indicators serve policy and the agricultural sector in identifying options for enhancing these beneficial effects of agriculture.

The second objective is to monitor the response of the agricultural sector to policy targets established for the environment, landscape and nature, in terms of adjustment of farming practice and farm management. Indicators identify the uptake of policy measures and support the evaluation of environmental targets, which are formulated in agricultural policies (e.g. the Nitrates Directive; agri-environmental measures).

Agri-environmental indicators contribute to requirements for monitoring the 1992 reform of the CAP and its impacts on the environment, and monitor progress on directives that affect European agriculture (e.g. the Nitrates

©CAB INTERNATIONAL 1999. *Environmental Indicators and Agricultural Policy* (eds F.M. Brouwer and J.R. Crabtree)

Directive, Birds and Habitats Directive). An important issue is to link such indicators with the state of the environment, in terms of the quality of air, water, land and living resources (state indicators).

Several institutional arrangements have been established during recent years which contribute to the monitoring role of agri-environmental indicators:

1. The OECD Joint Working Party of the Committee for Agriculture and the Environment Policy Committee established a set of key agri-environmental indicators, exploring consistency of definitions and methods of measurement. In total some 13 indicators were proposed to address agri-environmental issues of relevance to policy makers (OECD, 1997; see also Chapter 3 of this volume).

2. The European Environment Agency (EEA) plays a major role in the provision of comparable and reliable information about the state of the environment at European level. A report on the state of the environment is published every 3 years (e.g. CEC, 1992a; EEA, 1995; EUROSTAT, 1995).

3. EUROSTAT (Statistical Office of the European Communities) instituted a programme with the aim of producing a European System of Environmental Pressure Indicators (ESEPI). This work addresses the target sectors of the Fifth Environmental Action Programme (energy, agriculture, transport, industry, tourism and waste). In addition, EUROSTAT initiated several efforts to operationalize indicators. Indicators are proposed describing pressures on the environment resulting from human activities for ten policy fields defined in the Fifth Environmental Action Programme (CEC, 1992b). These indicators are ranked according to policy relevance, analytical soundness and response elasticity. Some examples of biodiversity indicators are presented in Chapter 4 of this volume.

Recent progress has been achieved in the monitoring of use and treatment of inputs by European agriculture. Well-accepted approaches have been developed to assess air emissions from societal activities (including agriculture). Various indicators are available for water pollution and water resources. Methodologies have been explored to develop indicators linked to nutrient balances (e.g. EUROSTAT, 1997), initiatives have been taken to develop indicators in the field of pesticides (OECD, 1997) and data are available on the abstraction of water by agriculture (EEA, 1995). However, major efforts are still required to monitor the effects of the CAP on landscape and nature.

The objective of the present chapter is first to review policy requirements for agri-environmental indicators, and secondly to identify data availability for making such indicators operational in the context of the European Union (EU). A distinction is made between data collated by international organizations (e.g. EEA, EUROSTAT, OECD) and that provided by Member States. Some conclusions with respect to data requirements and data availability are made in the final section.

POLICY REQUIREMENTS IN THE EUROPEAN UNION

Environmental requirements increasingly place conditions on agriculture in the EU. Such requirements may result from agricultural policies as well as from environmental legislation. For example, in the 1992 CAP reform, measures were adopted to improve the environmental soundness of agricultural production. Environmental requirements are currently included in Council Regulations on products that include arable crops, beef and sheep. Also, environmental conditions are imposed on the Less Favoured Area Scheme, and on programmes under Objectives 1 and 5b (Brouwer and Van Berkum, 1996).

Concern about the negative effects of agriculture on the environment was voiced at the official level during the mid-1980s. The EC's Third Action Programme for the Environment, published in 1983, stated the need to

> 'promote the creation of an overall strategy, making environmental policy a part of economic and social development, [resulting] in a greater awareness of the environmental dimension, notably in the field of agriculture [and] enhance the positive and reduce the negative effects on the environment of agriculture . . .'

> (*Official Journal of the European Communities*, C.46, 17 February 1983).

The Fifth EC Environmental Action Programme *Towards Sustainability* aims to achieve ecologically sustainable economic development. Agriculture is one of the five target sectors selected by the Commission for special attention, the others being industry, energy, transport and tourism. The target sectors chosen are those where a Community approach 'is the most efficient level at which to tackle the problems these sectors cause or face' (CEC, 1992b, p. 6).

Targets and actions required in agriculture are formulated in the Fifth Environmental Action Programme. A list of these targets for the year 2000 and actions needed in relation to the agriculture and forestry sectors are given in Table 5.1. No specific targets are included to monitor the development of rural areas of the Community and the sector's productive, social and environmental functions (CEC, 1992b, p. 7).

The following environmental policy objectives are most relevant to the agricultural sector:

The Nitrates Directive

This directive concerning the protection of waters against pollution caused by nitrates from agricultural sources was issued by the Council in December 1991 (91/676/EEC) (*Official Journal of the European Communities*, 1991). It is under the responsibility of the Directorate-General Environment, Nuclear Safety and Civil Protection (DG XI). Policies are being formulated in several Member States in order to reduce pollution of groundwater (nitrates), surface water

Table 5.1. Medium-term environmental targets up to the year 2000 and actions needed in agriculture (source: CEC, 1992b, p. 37).

Targets up to the year 2000	Actions
Standstill or reduction of nitrate levels in groundwater	Strict application of the Nitrates Directive
Reduced incidence of surface waters with a nitrate content exceeding 50 mg l^{-1} or giving rise to eutrophication of lakes and seas	Setting of regional emission standards for new livestock units (ammonia) and silos (silage)
	Reduction programme for phosphate use
Stabilization of increase of organic material levels in the soil	Allocation of premiums and other compensation payments to be subject to full compliance with environmental legislation
Significant reduction of the use of plant protection products per unit of land under production and conversion of farmers to methods of integrated pest control, at least in all areas of importance for nature conservation	Registration of sales of plant protection products
	Registration of use of plant protection products
	Control on sale and use of plant protection products
	Promotion of 'integrated control' (in particular training activities) and promotion of bio-agriculture
Fifteen per cent of agricultural area under management contracts	Programmes for agriculture/ environmental zones with premiums co-financed by EAGGF
	Protection of all endangered domestic animal races
	Programmes for agriculture/environment
Management plans for all rural areas in danger	Re-evaluation of licence conditions for irrigation and of state aids for drainage schemes
	Training of farmers, promotion of exchange visits between regions with comparable environment management situations
Increase of forest plantation, including on agricultural land	New afforestation and regeneration of existing forest, favouring the most adequate means for the environment (slow-growing trees, mixed afforestation)
Improved protection (health and forest-fires)	Further action against forest-fires

(eutrophication by excessive use of nitrogen and phosphate fertilizers) and the atmosphere (ammonia emissions). Directive 91/676 includes regulations on how to handle manure and fertilizers in zones which are identified as vulnerable to the leaching of nitrate. Member States need to establish action programmes with rules for treating nutrients from organic and inorganic sources. The application of nutrients should comply with Codes of Good Agricultural Practice. Monitoring programmes are required to assess the effectiveness of the action programmes based on the quality of the aquatic environment. Livestock manure is only part of the input flows through agriculture and an insufficient indicator of the potential of leaching of nitrates to water. Nitrogen balances are more appropriate tools for providing insights into the balance between input and output flows.

Integrated Pollution Prevention and Control (IPPC)

A common position was adopted by the Council on 27 November 1995 with a view to adopting a Council Directive concerning integrated pollution prevention and control (*Official Journal of the European Communities*, 25.3.96) (96/C 87/02) (Directive on Integrated Pollution Prevention and Control). The purpose of this directive is to achieve integrated prevention and control of pollution arising from the activities listed in Annex I of the directive. It applies to certain installations in the energy industries, the production and processing of metals, the mineral industry, the chemical industry, waste management and other activities. Regarding agricultural activities, the directive also applies to installations for the intensive rearing of poultry and pigs.

Environmental Impact Assessment Directive (EIA)

The European Commission recently presented a proposal for a directive to amend Directive 85/337 (EEC) on the assessment of the effects of certain public and private projects on the environment. The amendments would considerably increase the number of impact assessments required compared with the 1985 Directive. The directive is to ensure pollution prevention and fair competition between producers on the internal market.

Habitats Directive

Council Directive 92/43 on the conservation of natural habitats and of wild fauna and flora (Habitats Directive) is 'to contribute towards ensuring biodiversity through the conservation of natural habitats and of wild fauna and flora of Community interest'. Member states need to communicate a list of sites to be designated as protected areas constituting the Natura 2000 network. It is a coherent network of special areas of conservation and includes the Special Protection Areas (SPAs) classified under the Birds Directive (79/409/EEC) and the Special Areas of Conservation (SACs) to be designated under the Habitats Directive.

North Sea Conference

Monitoring programmes and data collection efforts are required to meet the objectives of the North Sea Conference which aim to reduce the nutrient load of coastal and marine waters. A scheme for reducing nutrient and heavy metal levels was agreed by the Ministers of the countries surrounding the North Sea, in which losses of nitrogen and phosphorus were to be reduced by some 50% during the period 1985–1995. The Working Group on Nutrients (NUT) of the Oslo and Paris Commissions provides biannual reports on Nutrients in the Convention Area: Overview of Implementation of PARCOM Recommendation 88/2 (Oslo and Paris Commissions, 1995). The report provides national data for application rates of fertilizers as well as livestock manure and mineral surplus calculations.

DATA AVAILABILITY IN THE EUROPEAN UNION

Efforts by International Organizations

Data are available from various sources to contribute to the operationalization of agri-environmental indicators. Data availability is reviewed in this section, based on information provided by the European Environment Agency (EEA), EUROSTAT, OECD and the Oslo and Paris Commissions.

European Environment Agency
The CORINE (co-ordination of information on the environment) inventories were initiated in the mid-1980s to provide information on the state of Europe's environment and natural resources. Methodologies and databases have been established since then, specifically in the field of land cover, biotopes and air emissions. The CORINE system also includes information on natural resources and the state of forests, as well as the quality of bathing water and fresh water.

The objective of the EEA is:

to provide the Community and the Member States with:

- objective, reliable and comparable information at the European level enabling them to take the requisite measures to protect the environment, to assess the results of such measures and to assure that the public is properly informed about the state of the environment;

- to that end, [to provide] the necessary technical and scientific support.

(Article 1 of Regulation 1210/90)

The EEA reports every 3 years on the state of the environment in the EU (EEA, 1995). EEA and a wide range of research organizations across Europe embody the European environment information and observation network (Eionet), including European topic centres, national reference centres and national focal points.

OECD

The OECD Group on the State of the Environment (SOE) is developing a core set of environmental indicators across all sectors of the economy. The SOE regularly provides information to feed into the *OECD Environmental Data Compendium*, as well as the core set of environmental indicators – *Environmental indicators – OECD Core Set*. Indicators related to the agriculture sector refer to agricultural activities that generate pressures on the environment, such as pollution by nitrogenous and phosphate fertilizers, as well as pesticides. Indicators distinguished are (OECD, 1995):

- Land used for agriculture, specifically arable and permanent crop land (source: FAO, OECD).
- Land used by agriculture, as permanent grassland (source: FAO, OECD, national statistical yearbooks).
- Irrigated area for agricultural purposes. Data relate to areas purposely provided with water, including land flooded by river water for crop production or pasture improvement (controlled flooding), whether this area is irrigated several times or only once during the year (source: FAO, OECD).
- The economically active population in the primary sector, comprising persons engaged principally in agriculture, forestry, hunting or fishing (source: OECD, FAO).
- Mechanization of agriculture, notably the use of tractors and combined harvester-threshers. The growth in the stock of tractors and other machinery is considered to be an important aspect of farming practices and a factor in explaining the impact of farming on the environment (source: FAO).
- Final consumption of commercial energy sources by the agricultural sector, exclusive of non-commercial sources of energy (source: OECD, International Energy Agency).
- Consumption of commercial nitrogenous fertilizers by agriculture (source: FAO, International Fertilizer Industry Association (IFA)).
- Consumption of commercial phosphate fertilizers by agriculture (source: FAO, IFA).
- Consumption of agricultural pesticides, including total pesticides, insecticides, fungicides and herbicides (source: FAO, national statistical

yearbooks, United Nations Economic Commission for Europe (UNECE), United Nations Environment Programme (UNEP)).

- Numbers of selected domestic animals: cattle, sheep and goats, horses, mules and asses, and pigs (FAO).
- Volume of aggregate agricultural production (source: OECD, FAO).

Environmental indicators distinguished by environmental theme are also presented in OECD (1994). A distinction is made between: (i) climate change and ozone layer depletion, (ii) eutrophication, (iii) acidification, (iv) toxic contamination, (v) urban environmental quality, (vi) biodiversity and landscapes, (vii) waste and (viii) natural resources.

Availability by Environmental Theme

Data availability differs largely along the various policy fields of the Fifth Environmental Action Programme (e.g. climate change, ozone layer depletion, loss of biodiversity, resource depletion, waste, air pollution, dispersion of toxics, water pollution and water resources, marine environments and coastal zones). Information regarding some indicators for air pollution, climate change, toxics and waste is collected on a routine basis for all member states, and classified according to economic activities. The Intergovernmental Panel on Climate Change (IPCC), for example, is currently initiating major data collection efforts. CORINAIR and the efforts of the IPCC will be used in the Environmental Pressure Information System (EPIS) of EUROSTAT. A more detailed review of indicators for the agriculture sector is provided in Bittermann *et al.* (1998).

Climate change, air pollution and ozone layer depletion
A proper methodology to quantify emissions across economic sectors is essential. Coefficients are used in order to assess emissions by economic activity (including agriculture). A standard nomenclature (SNAP) is applied in agriculture, forestry, land use and change in timber stocks (SNAP 10). Close cooperation exists between the UN–ECE and EEA, as well as the IPCC (e.g. on emissions of methane (CH_4) and nitrous oxide (N_2O)). Substances covered by the Montreal Framework, e.g. methyl bromide, are not covered by the European Topic Centre (ETC) on Air Emissions.

Different definitions exist regarding emissions of carbon dioxide (CO_2). First, differences might arise due to the definition of the territory of a country and due to the national responsibility. Consider, for example, the role of 'shipping' and 'air traffic'. IPCC, for example, takes the domestic position and excludes emissions related to international transport. Such differences need to be considered in further efforts on data availability and data requirements.

Emission inventories made so far are based on the year 1985 (12 countries and three pollutants: VOC (volatile organic compounds), SO_2 and NO_x), 1990 (29 countries and 12 pollutants, including acidifying substances and greenhouse gases) and 1994 (30 countries and 20 pollutants) (European Topic Centre on Air Emissions). Only a few countries provided estimates for emissions of heavy metals and persistent organic pollutants (POPs) and, therefore, results are derived from the expert panel on POPs. These sources of pollution are certainly not applicable to the full coverage of the EU, but need to be seen as an indication.

The Corinair project (core emission inventory AIR) is part of the work programme of the EEA (ETC on Air Emissions, 1997). It was developed to cover all relevant sources of emissions to air of a number of pollutants, relevant for several environmental themes. In total 11 main sources of pollution have been distinguished (SNAP, level 1), including agriculture, forestry, land use and the change in timber stocks as one activity. The total number of sub-groups distinguished (Selected Nomenclature for emissions of Air Pollution, SNAP, level 3) is presently around 400 (Jol and McInnes, 1997).

The ETC on Air Emissions in 1996 performed a spatially detailed (NUTS 3 level) air emission inventory for 1994 for sulphur dioxide (SO_2), nitrogen oxides (NO_x), non-methane volatile organic compounds (NMVOC), ammonia (NH_3), carbon monoxide (CO), methane, nitrous oxide and carbon dioxide. In addition, the inventory also includes nine heavy metals and ten POPs, including dioxins and furans. Work of the ETC/AEM was developed in cooperation with the UN–ECE Task Force on Emission Inventories, and is consistent with international energy statistics and the OECD/IPCC guidelines for emission inventories for greenhouse gases (Jol and McInnes, 1997). Air emission inventories for the years 1995 and 1996 are expected to be available before the end of 1998. Countries have committed themselves to supply greenhouse gas data according to the IPCC and data on emissions of SO_2, NO_x, NH_3 and NMVOC according to the European Monitoring and Evaluation Programme (EMEP).

Water pollution and water resources
An assessment of the environmental state of European rivers and lakes was prepared by the National Environmental Research Institute (NERI) of the Danish Ministry of Environment and Energy to serve as a background document to the Dobříš Report (Kristensen and Hansen, 1994). According to this report, phosphorus and ammonium concentrations in most of the European rivers and many lakes have improved over the past 10–15 years. Nitrate levels, however, increased during that period, mainly due to the increasing use of fertilizers. Nitrate levels increased during the periods 1977–1982 and 1988–1990 in more than two-thirds of the 230 European rivers investigated.

Nutrient balances
Nutrient balances are indicators for the environmental themes of resource depletion, water pollution and water resources. They also contribute to eutrophication, which is a major source of pollution of the marine environment and of coastal zones. EUROSTAT is making progress in the operationalization of

nutrient balances. The approach used is based on the surface balance concept (Brouwer *et al.*, 1995), although difficulties remain with availability of information. It is recommended that nutrient balances be quantified at the regional level, focusing on specific sensitive areas. Inputs of data on mineral fertilizers are based on FAO statistics at national level but no data are available describing the use of mineral fertilizers at a regional level. Inputs due to livestock effluents are based on coefficients regarding the nitrogen content in manure. Difficulties arise with the assessment of mineral uptake by crops, for example when both total production and utilization of pastures is assessed (in dry matter terms). In addition, there are problems with the assessment of the total production of pastures, as this differs between harvested and grazed pastures; non-utilized pastures are not required for the calculation of nutrient balances.

The work of EUROSTAT is based on data from the Farm Structure Survey for the year 1993, and includes both input and output flows. The methodology is described in OECD (1996). Figures are now available from OECD for a period of 10 years.

Crop protection products

Agricultural use of crop protection products is considered a major contributor to the harmful effects of agriculture on the environment. Usage of crop protection products and their emissions to the aquatic environment are considered to be important indicators to loss of biodiversity and dispersion of toxins as well as water pollution and water resources. Figures on sales of crop protection products in Europe in the period 1985–1995 are presented in ECPA (1996). Data collected in that report include the volume of active ingredients sold (tonnes) in the EU-15 (with the exception of Luxembourg) and differentiate between herbicides, insecticides, fungicides and others. Such figures enable the comparison of sales of crop protection products among Member States. Two important things, however, need to be considered when using and interpreting this information:

- The report is based on information from national industry associations on sales of crop protection products (in tonnes of active ingredients). Such figures are estimated to cover around 90% of the European market.
- Statistics on sales may differ and not accurately reflect the use of these products in agriculture, as farmers may keep stocks.

A limited number of EU member states, including the United Kingdom and The Netherlands, do have a detailed database on use of crop protection products. However, the total amount of crop protection products used is an imperfect indicator of soil and water quality. Difficulties would therefore arise in linking pesticide use levels with quality of the European environment (e.g. pesticide loading in water). Possible agri-environmental indicators could be:

- Total area treated. This may show an increasing trend even when the total amount used is declining.

- Relative environmental loading (REL). This has decreased consistently since 1988.
- Ecotoxicological loading, which was not previously considered.

Water use for irrigation purposes

The amount of water used by the agricultural sector is an important indicator of water pollution and water resources. Large differences exist between member states regarding the amount of water used agriculturally, the share of water used for irrigation purposes in total national water consumption being less than 10% in Austria and over 80% in Spain. Statistics on water consumption are available from the International Water Supply Association.

Loss of biodiversity and resource depletion

Environmental indicators related to (the loss of) biodiversity are as yet hardly defined. With regard to user needs and data availability, this is one of the most difficult pressure indicators to be put into operation. Basic improvements are still needed, particularly in terms of defining objectives and developing methodologies for measuring change in biodiversity. Different scales need to be considered, ranging from local to biogeographic. Also, soil erosion may have to be operationalized in the context of loss of biodiversity, and a broader investigation on soil quality may be essential, possibly under the heading of resource depletion. Rather than focusing on geographical scales (e.g. countries, regions) it is recommended that the focus should be on biogeographic regions (e.g. alpine regions). Naturalness could be one of the criteria to be considered in an assessment of loss of biodiversity.

A database is likely to become available in the next few years in response to the flora and fauna Habitats Directive.

Pressures on farmland and wildlife may potentially increase in response to:

- increased use of pesticides and of fertilizers;
- higher stocking rates for livestock and livestock numbers;
- greater specialization and loss of mixed farming;
- shift from spring to autumn sowing of crops;
- shift from hay to silage production;
- changes in crop rotations and loss of fallows;
- increased irrigation and drainage;
- increased field size and loss of hedgerows;
- loss of agricultural land through urbanization, abandonment and afforestation.

The nature conservation status of birds was reviewed by Birdlife International in 1994 (Tucker and Heath, 1994). The increasing intensity of agriculture is considered to be one of the greatest threats to the bird population, including use of crop protection products, as well as agricultural abandonment in some areas.

Indicators on landscape may cover the area of land designated for protection. The extent of environmentally managed land, which entered into Regulation 2078/92, could be a first step towards indicators for landscape.

DATA AVAILABILITY FROM MEMBER STATES

Several Member States have increased substantially their efforts to assess the impact of changes in policy on the environment. The concept of pressure–state–response relationships is being used increasingly. Table 5.2 indicates national sources of data and long-term trends from environmental policy reviews.

Belgium

Verbruggen (1996) reports on the state of the environment and nature for the Flanders region. Linkages among societal activities and pressure on the environment are included. A distinction is made between societal activities and development trends regarding population, industry, agriculture, transport, energy and water.

The Netherlands

Similar reports on the state of the environment are provided at regular intervals in The Netherlands (RIVM, 1997). Environment balances for The Netherlands are published annually, and aim to present the information that is required to contribute to the monitoring of all environmental policies in that country. A distinction is made between monitoring of:

Table 5.2. National documents providing data and long-term trends.

Member states	Source
Belgium	Verbruggen (1996)
Denmark	Christensen *et al.* (1994)
Germany	Bundesministerium für Umwelt, Naturschutz und Reaktorsicherheit (1994)
Finland	Puolamaa *et al.* (1996), Rosenström *et al.* (1996)
Luxembourg	Ministère de l'Environnement (1994)
Netherlands	RIVM (1997)
Austria	STAT/Umweltbundesamt (1994)
United Kingdom	HMSO (1994)

- policy – which includes a review of objectives, strategies, instruments and priorities, and it also reviews policy measures;
- target groups – which focuses on societal activities (including agriculture) and their impact on the environment, and also includes environmental pressures from agriculture; and
- environmental quality – which focuses on the biotic and abiotic environment and their effects on people, health, safety and ecosystems.

Austria

The report STAT/Umweltbundesamt (1994) depicts the state of the environment in Austria. It links activities (population, energy, agriculture, industry and waste) with pressure on the environment (emissions, water, soil, forests, nature and landscape) and societal response (environmental policy measures taken and expenditures to environmental policy). Indicators relating to agriculture include livestock manure, ammonia emissions, biodiversity, soil degradation and pesticides.

Denmark

Christensen *et al.* (1994) depicted the state of the environment and nature in Denmark, and linked trends with the developments in human activity that had the greatest impact on the environment. The state of the environment assessment included air pollution, urban environment, eutrophication, groundwater, biodiversity and environmentally hazardous substances. Societal activities that impact on the environment include economic development, energy, transport, agriculture and forestry, industry and households. Agricultural indicators include intensity of agriculture (e.g. livestock density and arable crop production), use of fertilizers and pesticides and leaching of nitrogen from agricultural land. It is intended that the status report be updated every 4 years.

Finland

The report by Rosenström *et al.* (1996) provides indicators for the 1997 OECD Environmental Performance Review of Finland. These indicators are presented according to the pressure–state–response framework applied in the Core Set of OECD indicators. A limited number of indicators directly link pressure on the environment with agriculture. Pressure indicators for eutrophication include phosphorus and nitrogen from livestock and fertilizer use, and nutrient balances in agriculture. The main indicator for toxic contamination of importance to the agricultural sector is pesticide use. Changes in agricultural practices are considered to be one of the threats to species survival.

In the future, work in Finland on environmental indicators will be integrated with their efforts to contribute to the United Nations Commission for Sustainable Development. A report on the key indicators for sustainable development in Finland will be published biannually.

CONCLUSIONS

1. Information requirements for agri-environmental indicators focus on the use and treatment of inputs by the agricultural sector, their emissions to the environment, and the state of the environment. There is an increasing need for so-called integrated environmental impact assessments to assess the relationships between environmental problems and the causative factors in society (including agriculture). Efforts are required to integrate pressure indicators with those covering the 'state' and 'response' or 'driving force' measures.

2. Data on use of inputs by the agricultural sector are available on a regular basis at member states level, including fertilizers (FAO), crop protection products (ECPA) and water for irrigation purposes (International Water Supply Association). Methods for assessment of nutrient balances at national and regional level have improved during recent years, and assessments have been made by OECD and EUROSTAT.

3. Data quality varies. Data availability is good for specific areas (e.g. air pollution and climate change) but more restricted for other environmental themes (e.g. biodiversity). Loss of biodiversity and landscape are the main areas for further work. There is a need for agreement regarding the types of pressures to be included, as well as the data gaps and data availability.

4. Indicators need to be operationalized at sufficient spatial differentiation in order to reflect policy needs at regional level. Figures at national level may be insufficient to assess the beneficial effects of CAP reform measures on the environment, landscape and nature.

5. Important criteria for data collection are completeness, consistency and comparability between countries, transparency to allow construction of a database from national inventories, and timeliness.

REFERENCES

Bittermann, W., Geissler, S., Brouwer, F. and Hellegers, P. (1998) *Joint Final Report of SIP Agriculture*. Statistical Office of the European Communities, Luxembourg (forthcoming).

Brouwer, F.M. (1995) *Indicators to Monitor Agri-environmental Policy in the Netherlands*. Mededeling 528, Agricultural Economics Research Institute (LEI-DLO), The Hague, The Netherlands.

Brouwer, F.M. and Van Berkum, S. (1996) *CAP and Environment in the European Union: Analysis of the Effects of the CAP on the Environment and Assessment of Existing Environmental Conditions in Policy*. Wageningen Pers., Wageningen, The Netherlands.

Brouwer, F.M., Godeschalk, F.E., Hellegers, P.J.G.J. and Kelholt, H.J. (1995) *Mineral Balances at Farm Level in the European Union*. Onderzoekverslag 137, Agricultural Economics Research Institute (LEI-DLO), The Hague, The Netherlands.

Bundesministerium für Umwelt, Naturschutz und Reaktorsicherheit (1994) Umweltpolitik: Umwelt 1994. Politik für eine nachhaltige, umweltgerechte Entwicklung. Bonn, *Deutscher Bundestag*, 12. Wahlperiode Drucksache 12/8451. 06.09.94

CEC (1992a) *The State of the Environment in the European Community: Overview*. COM(92) 23 final – Vol. III, 27 March 1992. Commission of the European Communities, Brussels.

CEC (1992b) *Towards Sustainability: A European Community Programme for Policy and Action in Relation to the Environment and Sustainable Development*. Commission of the European Communities, Brussels.

Christensen, N., Paaby, H. and Holten-Andersen, J. (eds) (1994) *Environment and Society – A Review of Environmental Development in Denmark*. Ministry of the Environment, National Environmental Research Institute, Department of Policy Analysis, Roskilde, Denmark.

ECPA (1996) *European Crop Protection: Trends in Volumes Sold, 1985–1995*. European Crop Protection Association, Brussels.

EEA (1995) *Environment in the European Union 1995: Report for the Review of the Fifth Environmental Action Programme*. European Environment Agency in cooperation with EUROSTAT, Copenhagen.

ETC on Air Emissions (1997) *CORINAIR 94: Summary Report 1*, Report to the European Environment Agency from the European Topic Centre on Air Emissions. European Environment Agency, Copenhagen.

EUROSTAT (1995) *Statistical Compendium for the Dobríš Assessment*. Statistical Office of the European Communities, Luxembourg.

EUROSTAT (1997) *Soil Surface Nitrogen Balances in EU Countries*. Internal working document. Working group 'Statistics of the environment', Sub-group on nitrate balances. Joint Eurostat/EFTA Group, February 13–14, 1997. Statistical Office of the European Communities, Luxembourg.

HMSO (1994) *This Common Inheritance. The Third Year Report*. HMSO, London.

Jol, A. and McInnes, G. (1997) CORINAIR: Towards a European PRTR or Integrated Emission Inventory. Paper presented at Pollutant Release and Transfer Registers (PRTRs), Workshop for Central and Eastern Europe and the New Independent States of the former Soviet Union, 15–17 January 1997, Prague, Czech Republic.

Kristensen, P. and Hansen, H.O. (1994) *European Rivers and Lakes: Assessment of Their Environmental State*. EEA Environmental Monographs 1. European Environment Agency, Copenhagen.

Ministère de l'Environnement (1994) *L'état de l'Environnement 1993*. Ministère de l'Environnement, Luxembourg.

OECD (1994) *Environmental Indicators. OECD Core Set*. Organisation for Economic Co-operation and Development, Paris.

OECD (1995) *OECD Environmental Data. Compendium 1995*. Organisation for Economic Co-operation and Development, Paris.

OECD (1996) *Development of an Agricultural Nutrient Balance Indicator: Progress Report from Belgium*. Report to the Joint Working Party of the Committee for Agriculture and the Environment Policy Committee, June 19–21 1996. COM/AGR/CA/ENV/EPOC(6)84. Organisation for Economic Co-operation and Development, Paris.

OECD (1997) *Environmental Indicators for Agriculture*. Organisation for Economic Co-operation and Development, Paris.

Oslo and Paris Commissions (1995) *Nutrients in the Convention Area: Overview of Implementation of PARCOM Recommendation 88/2*. Oslo and Paris Commission, London.

Puolamaa, M., Kaplas, M. and Reinikainen, T. (1996) *Index of Environmental Friendliness: A Methodological Study*. Statistics Finland, Helsinki.

RIVM (1997) *Milieubalans 1997*. Alphen aan den Rijn, Samson H.D. Tjeenk Willink bv/Bilthoven, National Institute for Public Health and Environmental Protection.

Rosenström, U., Lehtonen, M. and Muurman, J. (1996) *Trends in the Finnish Environment. Indicators for the 1997 OECD Environmental Performance Review of Finland*. The Finnish Environment 63. Ministry of the Environment, Environment Policy Department, Helsinki.

STAT/Umweltbundesamt (1994) *Umwelt in Österreich. Daten und Trends 1994*. Österreichischen Statistischen Zentralamt und dem Umweltbundesamt, Vienna.

Tucker, G.M. and Heath, M.F. (1994) *Birds in Europe: Their Conservation Status*. Birdlife Conservation Series No. 3. Birdlife International, Cambridge, UK.

Verbruggen, A. (ed.) (1996) *Milieu- en natuurrapport Vlaanderen 1996: leren om te keren*. Vlaamse Milieumaatschappij en Garant Uitgevers, Leuven/Apeldoorn, The Netherlands.

Establishing Targets to Assess Agricultural Impacts on European Landscapes

6

Dirk M. Wascher, Marta Múgica and Hubert Gulinck

INTRODUCTION

Today, there is increasing awareness that many of the environmental problems affecting European landscapes originate outside the region concerned. A global market economy, the impacts of the Common Agricultural Policy, trans-European traffic networks, large-scale demographic and socio-economic change, cross-boundary (e.g. airborne) pollution, as well as differences in landscape-related policy and planning mechanisms at the national level, have left European landscapes exposed to unprecedented trends and threats. Despite the growing international appreciation of Europe's characteristic landscapes, regional or national capacities to detect, let alone to prevent and counteract the observed changes, have clearly proved to be limited. Most efforts have been directed at traditional nature conservation, resulting in a substantial increase in nationally and internationally protected areas in Europe. While this must certainly be considered as a success in its own right, the continually increasing number of endangered plants and animal species brings the adequacy of these measures into question. Site and species protection must stand isolated if surrounding landscapes cannot provide the necessary ecological infrastructure by connecting fragmented and sensitive core habitats to larger ecological units and biological life-support systems. Awareness of the level of ongoing landscape change and the possibilities for countervailing measures is sometimes limited to a very few expert groups. It is not surprising, therefore, that Europeans are witnessing the loss or degeneration of traditional and characteristic landscapes in their native regions.

European institutions have only lately begun to respond to these problems. In 1990, concern for the state of landscapes led the International

©CAB INTERNATIONAL 1999. *Environmental Indicators and Agricultural Policy* (eds F.M. Brouwer and J.R. Crabtree)

Association of Landscape Ecologists (IALE) and the World Conservation Union (IUCN) to establish a Landscape Conservation Working Group (LCWG) as part of the Commission on Environment (CESP). Since its establishment in 1978, CESP's objective has been to identify, document and safeguard threatened landscapes throughout the world. At the strictly European level, landscapes have become the object of environmental reporting (Council of Europe, 1995; Stanners and Bourdeau, 1995). Parallel to this, a European Landscape Convention is being developed.

For decades, landscape ecological disciplines have provided the methodological and conceptual framework for a wide arena of environmental research and planning. Hardly any urban or rural development plan, road construction, river course adjustment or other large-scale physical planning task has been executed without an input from the discipline of landscape ecology. Following the introduction of 'environmental impact assessment' in the early 1980s, the profession's technical instruments and concepts have become increasingly refined in order to provide decision-makers and the public with objective and reliable tools. While landscape ecology has become a technically advanced and permanently applied component of land planning and object (impact)-orientated decision-making processes, the discipline is limited to being a rather 'reactive' tool at the regional and local level. At the supra-regional and national levels, planning schemes continue to be rooted in regional planning as the mitigating instrument to balance different societal interests, such as agriculture, industry, urban development and *nature and landscape conservation*. One of the few exceptions is the role of 'the ecological main structure' as it has been laid down in the Nature Policy Plan of The Netherlands (MANMF, 1990).

STATE INDICATORS FOR LANDSCAPE VALUES AND FUNCTIONS

The assessment of biological and landscape diversity has recently become a topic of research among landscape ecologists as well as environmental economists. The valuation of landscapes, certainly at the European level, is a very complex matter for which no standard methods have yet been developed. Studies to date have been rather patchy and inconsistent, hindering their use as a reference for European-wide landscape valuation. Table 6.1 provides brief definitions for the terms used in this chapter.

Three assumptions have been made for which value categories have been proposed: (i) all landscapes have a value; (ii) value categories can be recognized; and (iii) landscape valuation is an integration of perceptual, ecological, cultural and economic aspects (Gulinck and Múgica, 1996). The categories defined aesthetic, socioeconomic and ecological criteria. Based on these categories, value-ranking orders can be performed, although it is recognized that such a ranking must result from a combination of professional judgement, informed opinion and public preferences.

Table 6.1. Definition of terms for landscape assessment.

Classes or categories	Broad landscape 'paradigms' pertaining to the fields of perception, cultural history, ecology and economy (sustainable land use)
Criteria	Functional, qualitative, generic characteristics of landscape that can be deduced from the different landscape theories and that support the description of landscape values (e.g. aesthetic, socioeconomic and ecological)
Indicators	Objective dimensions of landscape criteria (e.g. connectivity, heterogeneity, spatial correlation, visual balance)
Indices or parameters	Mathematical formulations of indicators, such as Shannon index. They should be directly applicable on the data sets. For strictly qualitative/subjective criteria judgement scores should apply, or statistics deduced from public perception study

As the term 'landscape functions' has become more established in recent years, it is necessary to stress divergence from the notion of 'function' as compared with the use of the term in mathematics and politics. Landscape functions can generally be classified according to the main groups of production (economic), living space (social) and regulatory (ecological) processes (Bastian, 1991). In order to demonstrate the wide range of environmental and social aspects associated with the landscape approach, Bastian's landscape functions are presented as Table 6.2.

Other than regional or national scientific approaches such as that of Bastian, the study undertaken on behalf of the European Environment Agency (EEA) (1996) needed to ensure a European-wide application and to develop a methodology for measuring quality changes at the European level. Building upon the findings of the Dobříš Report (Stanners and Bordeau), the quality assessments proposed in the EEA Report introduced the notion of landscape values as equivalent to landscape qualities. The term 'value' had been chosen to indicate the fact that, differing from its use in ecological science, the landscape concept is closely related to human preferences. Thus, with the assessment of landscape values, human components are addressed more directly than is usually the case in classic ecological disciplines. Furthermore, the required concept needs to offer a spatial framework that integrates national as well as international perspectives. Because of these methodological challenges, it was necessary to focus on a smaller number of landscape functions and values than those listed in Table 6.2. Since indicators for measuring landscape functions and qualities are still under development, the proposed concept is based on theoretical considerations, taking into account the results of case studies (at the national level) and existing European data.

Table 6.2.   Landscape functions (capacity, production and disposition)
differentiated according to functional groups, main functions and sub-functions
(after Bastian, 1991, 1996).

A. Production (economic) functions
    Availability of renewable resources
      Biomass production
        Plant biomass (agricultural suitability)
          Cropland
          Permanent grassland
          Special cultures
          Wood
        Animal biomass
          Game
          Edible fish
      Water retention
        Surface water
        Groundwater
    Availability of non-renewable resources
      Mineral resources, building materials
      Fossil fuels
B. Regulation (ecological) functions
    Regulation from material and energy circulation
      Pedological functions (soils)
        Soil protection against erosion
        Soil protection against flushing
        Soil protection against desiccation
        Soil capacity to actively break down disturbing factors (filter, buffer,
        transformation)
      Hydrological functions (water)
        Groundwater recharge capacity
        Water retention, discharge balance
        Self-cleaning capacity of surface water
      Meteorological functions (climate, air)
        Temperature balance
        Increasing air humidity, evaporation
        Windfield influence
    Regulation and regeneration of populations and biological systems
      Biotic reproduction and regeneration of bio-systems
      Regulation of organism populations (e.g. pest)
      Maintenance of genetic pool

Table 6.2. *contd.*

C. Living space (social) functions
    Psychological functions
        Aesthetic functions (landscape scenery)
        Ethnic functions (genetic pool, cultural heritage)
    Information function
        Function for research and education
        Biological indicators for state of the environment
    Human ecology functions
        Bio-climatic (meteorological) functions
        Filter and buffer functions for chemical processes
        Acoustic effects (noise protection)
        Recreational functions (complex of psychological and human ecological functions)

'GENERAL LANDSCAPE PROFILES' FOR INDIVIDUAL LANDSCAPE TYPES

As explained earlier, successful methodology for the assessment of landscapes is strongly dependent on the establishment of transparent and generally accepted reference values against which the measured criteria can be judged. For landscapes and biodiversity, such reference values are frequently defined for specific purposes at the regional or national level, but there is a lack of a coherent, operational and accepted landscape value system, especially at the international level. At the level of landscape planning for community and impact assessment purposes, landscape ecologists operate increasingly with the concept of specific objectives for a given landscapes, so-called 'general landscape profiles'. The general profile of a landscape is itself the result of a scientific analysis of its main physical and biotic properties with the goal of determining the optimal (ecologically sound, economically and socially beneficial) potential of a landscape. A general landscape profile is closely related to the term 'target profiles' which can be defined as: 'A description of the long-term, potentially desirable and basically feasible use, design and development of spaces or landscapes' (Brösse, 1975).

For this reason, dynamic ecological models have been developed to describe possible landscape changes in space and time, to simulate future developments and to explore compromise solutions by balancing conflicting land-use interests. It is not the intention with the development of general landscape profiles for landscapes to promote standardized or homogeneous solutions of the specific evaluation activities. The emphasis is on creating dynamic models that are open to various interpretations and which allow the traditional monopoly positions of some disciplines in providing one-dimensional definitions of our environment to be overcome. Account is thus taken of the views of agricultural as well as nature conservation experts. Figure 6.1

illustrates how target profiles can be developed, based on landscape functions at the regional level.

CONSOLIDATION PROCESS THROUGH EXPERT NETWORKS

Both the identification of regional landscape types as well as the development of general landscape profiles based on landscape functions require a high level of collaboration between international and regional experts. The bottom-up analysis of regional experts needs to meet and validate top-down criteria established by international teams. Ultimately, the development of general landscape profiles has to involve other disciplines and sectors to ensure a wide and long-term consensus in support of identified goals. The concept of general landscape profiles is very closely linked to the notion of 'social comprehension'. Such a process is also in line with Chapter 28 of Agenda 21 which calls explicitly for the establishment of regional networks to influence the decision-making process.

The following sections present the methodological and spatial framework in which indicator assessment could be developed, and identify suitable value categories as future tools for landscape assessment.

SPATIAL FRAMEWORK FOR LANDSCAPE INDICATORS

Any application of indicators for determining landscape functions and values is strongly linked to the question of spatial references for which such indicators can be considered as valid.

With large differences both in biophysical conditions throughout Europe, and the cultural and historical processes that have formed landscapes, it appears useful to identify areas of relative homogeneity as a reference base. Because the majority of existing regional and national concepts differ in terms of objectives and methodologies, there is a need for a common framework. The most important factors in such a framework are biogeographic parameters and land-use information recognized at the European level. This information can be used to produce generic landscape profiles (GLPs), i.e. areas of relative uniformity. The task will be facilitated by the existing information on biogeographic and ecological regions as well as by data from CORINE land cover (CEC, 1993). Both information layers will need to be verified by national and regional experts to ensure that existing information is taken into account before generic landscape profiles areas are defined (Fig. 6.2).

Each GLP area can be characterized by a set of biogeographic and land-use criteria. This set of criteria serves as the general framework for the actual landscape assessment using area-specific indicators and reference scales. The concept foresees the development of a spatial reference base in which the GLPs are broadly differentiated according to land-use intensity for rather broad

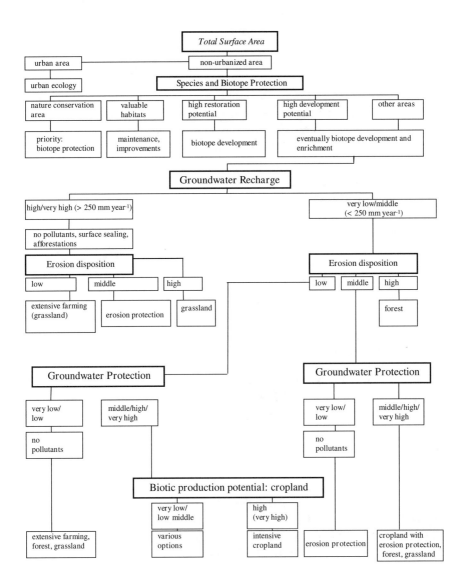

Fig. 6.1. Development of 'target profiles' at the regional level, based on landscape functions (after Bastian, 1996).

categories, such as rural, peri-urban and urban (Gulinck and Múgica, 1997). Distinctions between these categories can be based on geomorphological characteristics such as coasts, river lowlands or mountains.

Each regionally and nationally verified GLP can be described according to these broad categories. At this point, no landscape identification or

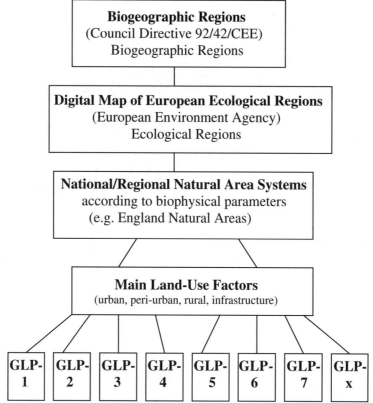

Fig. 6.2. Generic landscape profile (GLP) areas at the European level according to natural and anthropogenic conditions (reflecting mainly biogeographic and land-use aspects).

classification has taken place. This process remains reserved for a bottom-up expert consultation at national and regional levels (Fig. 6.3). However, these experts will have to follow a strictly defined procedure involving the parameters developed for the identification of the GLPs.

METHODOLOGICAL APPROACHES TO LANDSCAPE ASSESSMENT

Again, land use and, in particular, information on the intensity or management level of agricultural areas, will play a crucial role for this initial assessment (Table 6.2). The analysis of CORINE land cover and EUROSTAT data is likely to provide national and regional experts with a European view on the land-use intensity in specified areas. As a first step, the experts will be asked to verify this type of information, based on a set of rather coarse data. In doing so, they will also apply standard definitions of landscape types and elements as

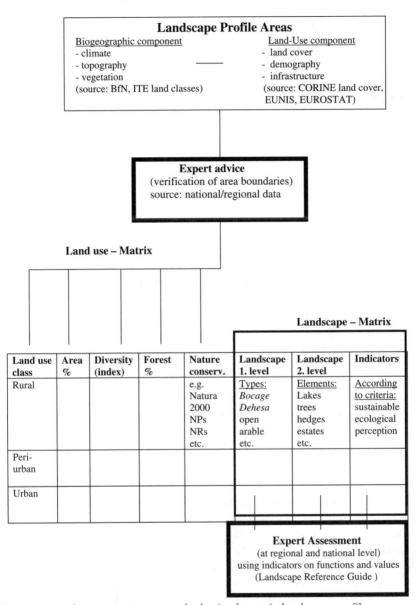

Fig. 6.3. Landscape assessment on the basis of generic landscape profile areas, applying agreed-upon indicators by national/regional experts.

these have been recognized at the European level. Independent of this verification process, the experts will also be asked to assess landscape functions and values in more detail by using indicators (see Table 6.3). According to Gulinck and Múgica (1997), an assessment based on landscape-specific indicators requires three main criteria: *sustainability, ecology* and *perception.*

Table 6.3. Table of proposed criteria and indicators.

Values related to major criteria	Indicators of complexity	Indicators of scale	Indicators of coherence	Indicators of composition	Indicators of stability (sustainability)
Aesthetic (perception)	Land cover diversity Colour diversity Relief energy Sequence of units Prospect/ refuge	Size of open spaces	Harmony (qualitative) Fragmentation indices	Water surfaces	Protection status Local supports
Ecological	Number of species/ area unit Biotope diversity	Size distri- bution of biotopes	Connectivity indicators Fragmentation indices	Valuable land-cover types Water surfaces	Disturbances Protection status Qualitative supply of water Regularity of water supply
Economic (land use)	Land-use diversity		Land use/soil correlation		Economic stability

Assessment procedures according to these three criteria can be considered as an extension of the identification process for GLPs: 'sustainability' through land-use information, 'ecology' through biogeographic information, and even 'perception' can use an analysis of land-use diversity based on CORINE data by means of the Shannon index (Magurran, 1988) as a first step to guide the process. Because of time and budget constraints, it has only been possible to develop a first overview on likely candidates for European indicators on land-scape assessment, and to test some of these for a case study area in Spain.

LANDSCAPE ASSESSMENT: SCALE, CRITERIA AND INDICATORS

The first step towards the definition of evaluation criteria is to consider three main value classes:

1. criteria related to the *perception component* (cultural and scenic features: the criteria that relate to the acceptance of landscapes by the general public and individuals);

2. the *structure and functional component* (ecological characteristics – in relation to the sustainability of biotic and abiotic features); and

3. the *land-use component* (mainly economic characteristics).

These three classes can be considered as sufficient to cover all major interpretations of landscape values as they exist in different professional groups in the field of landscape. Furthermore, there is a general agreement that the different criteria can be applied at different scale levels. With regard to scale, experts have repeatedly shown a preference for a 1 km^2 grid size as a spatial reference. It has the advantages of:

- matching the resolution of the CORINE land cover dataset (25 ha);
- matching the resolution for the Institute of Terrestrial Ecology (ITE) land classification system;
- being a good reference level for expressing the landscape value criteria and parameters;
- being a good reference level for upscaling (aggregation of km^2 cells) and downscaling (what constitutes the km^2).

Initially, and for certain areas, a much coarser resolution can be used. This depends also on the landscape context itself. For example, for peri-urban areas, a finer resolution should be adopted, whereas for very extensive agricultural open-field areas a coarse resolution may be suitable.

The identification of suitable indicators for the measurement of landscape values and functions at the European level is an ongoing process. Hence it is not possible to present indicators that have been internationally agreed or which are actually in use. Table 6.3 provides an overview of indicators according to to the main assessment criteria. Information on candidate indicators and relevant data sources is specified in Table 6.4. All of these candidate indicators need further testing and application at the European level.

CONCLUSIONS

A brief analysis of the existing assessment mechanisms for landscapes at the European level points at a number of conceptual and qualitative challenges with regard to data management in general and the use of indicators, or the development of implementation targets in particular. Recently, the implementation of pilot studies in the field of biological and landscape assessment at the European level has provided useful insights for future assessment strategies.

One of the key findings is that the technical part of landscape assessment by means of indicators needs to be supplemented by two crucial tasks:

Table 6.4.   Candidate indicators, data sources and linkages (Gulinck and Múgica, 1997).

Indicator	Data sources	Linkages
Land cover diversity	Shannon index; number of cover types km^{-2} Coarse resolution: CORINE Finer resolution: land cover sets derived from high-resolution satellite images Fine resolution: from airphotos, maps or field survey, including small landscape elements	Interaction with coherence
Architectural diversity	Number of construction styles km^{-2}: from field survey	
Colour diversity	Number of physiognomic vegetation types km^{-2}: from airphotos, land cover maps, vegetation maps, etc. Number of rock or soil types km^{-2}: from lithological/soil maps	
Seasonal diversity	Number of major phenological stages with distinct physiognomic expression: from climate map, land-use map, vegetation map (e.g. at least four stages in Belgium)	
Morphographic diversity	Number of land form types km^{-2}: from geomorphology maps, satellite images, contour maps Relief energy as max. altitude difference km^{-2}: from digital terrain model (DTM) or contour maps Number of basins of Nth order km^{-2}: from contour maps, stream maps	
Diversity of visual character zones	Number of distinct visual units: from field survey or from airphotos Number of transitions of character zones along roads Number of land-use types: from CORINE Number of remarkable focal elements or otherwise interesting visual features km^{-2}	Interaction with land cover and morpho-graphic diversity
Size of open spaces	Maximum view length from centre of km^{2} area comprised between visual barriers from detailed land cover map and specific spatial analysis software	
Size of biotopes	Number of biotope objects in different size ranges Size of total area of valuable biotopes Total edge length of biotopes	

Table 6.4. *contd.*

Indicator	Data sources	Linkages
Fragmentation	Length of major infrastructures km^{-2}: from road maps	
	Shannon index applied to groups of mutually unsupportive land uses: from interpreted land-use maps	
	Number of topographic objects km^{-2}: from detailed land-cover maps	
	Land-cover complexity index, e.g. through radial analysis in land-cover maps	
Purity of traditional land cover/ use	% land occupied by traditional land use	
	Wholeness of traditional pattern as e.g. alpha or beta network index for hedgerow landscapes	
	% abandoned land	
Connectivity	Aesthetic indices	
	Mean distance between biotopes	
	Inverse of fragmentation indices	
Land-use fit	Spatial correlation of land cover/soil type	
	% unfitted land cover	
Water surfaces	Diversity of surface water bodies as Shannon index applied to classification of water bodies from topographic maps	
	% area open water	
	Total shore length	
Water quality	Biotic index of water bodies	
	Distribution of water bodies in different quality classes	
	Contribution of unit area to groundwater recharge, yes/no	
	Natural water supportive to specific land uses, yes/no	
Soil quality	% area covered by highly fertile soils	
	Presence of lithological/geomorphologic/soil conditions important for nature conservation, yes/no	
	Concentration of remarkable objects of 'earth heritage', yes/no or objects km^{-2}	
Protection status	% area under protection rule	
	Number of protection rules applicable to unit area	
Land-use stability	Economic indicators of stability of interesting landscapes, e.g. agricultural succession safety, EC support, demography	
Ecological disturbances	% area prone to fire or inundation risk	
	Size affected by forest dieback, yes/no	
	% area affected by desiccation or desertification	

.en developing valid methodologies for landscape and biodiversity assessments, international experts need to establish procedures that link top-down approaches to national bottom-up activities. The development of implementation targets can only be successful if regional and national experts are provided with adequate assessment tools (aggregated, higher-level information) and when feedback and response mechanisms between international and national/regional experts occur early and are highly standardized throughout Europe.

2. Natural scientists, policy-makers and representatives of socio-economic sectors need to enter into an active dialogue to identify targets for future GLP areas. Without the definition of commonly accepted targets, without a scale to refer to, the decision-makers and the European public will not be able to interpret and accept indicators for biological and landscape diversity. A common understanding of definition for land use and landscape profiles can be considered as a priority objective for the integration of biodiversity into all sectors (see also Council of Europe, 1995).

Assessment procedures that lack the above components are not likely to meet the objectives because they will stand isolated or in contradiction to national and regional assessments.

In more specific terms, the following aspects appear to be of priority interest. First, by representing the largest share of Europe's environment, semi-natural and cultural landscapes are a stronghold of biodiversity. The landscape level appears to be the most appropriate context for such assessments since holistic, ecological and cultural principles are inherent to the very notion of it. With the help of a typology of European landscapes and standard methodologies, an attempt can be made to characterize any given landscape type in a comparable way. The scale of the landscape type in question can vary depending on the eco-regional conditions and project definitions.

Secondly, future assessments should be based on GLP areas rather than on administrative boundaries. From the methodological point of view, the existing European and national spatial reference schemes (e.g. the administrative NUTS2 regions) must be considered as inadequate for capturing essential environmental properties. The resource value of a region can only be assessed in the context of its biogeography *in combination* with supra-regional and European perspectives. The international dimension needs to be introduced into the regional level and, in return, regional aspects must find their way into international decision procedures. In order to calibrate the assessment instruments, it is proposed to use 'ecological regions' and 'landscape units' (both currently developed at the European level) as spatial reference units for assessments. In this way the availability or rareness of natural resources such as groundwater, soils, minerals, wood, but also of natural habitat with species of flora, as well as amenities such as cultural, recreational and aesthetic values, can be determined on the basis of common criteria which take the European perspective into account.

Finally, there is need for GLPs that define the region-specific objectives for sustainable land-use management on the basis of landscape criteria and indicators.

NOTE

Parts of this paper build upon the the contributions of Marta Múgica (Centro des Investigaciones Ambientales de la Comunidad de Madrid, Spain) and Hubert Gulinck (Catholic University Leuven, Belgium) to the EEA report *European Landscapes – Classification, Evaluation and Conservation*, D. Wascher (ed.), 1996, in press.

REFERENCES

Bastian, O. (1991) *Biotische Komponenten in der Landschaftsforschung und -planung. Probleme ihrer Erfassung und Bewertung.* Habitilationsschrift, Martin-Luther-Universität Halle-Wittenberg, Germany.

Bastian, O. (1996) Bestimmung von Landschaftsfunktionen als Beitrag zur Leitbildentwicklung. Beitrag zum BTUC Workshop Die Leitbildmehode als Planungsmethode. *BTUC Aktuelle Reihe.* 8/96, pp. 67–79.

Brösse, U. (1975) *Raumordnungspolitik als Entwicklungspolitik.* Schriftenreihe der Wissenschaftlichen Hochschule, Göttingen, Germany.

CEC (1993) *CORINE land cover. Technical Guide. Report EUR 12585.* European Commission, DG XI, Brussels.

Council of Europe (1995) *Pan-European Biological and Landscape Diversity Strategy.* Submitted by the Council of Europe at the Ministerial Conference Environment for Europe, Sofia, Bulgaria, 23–25 October 1995. ECE/CEP/23.

EEA (1996) European landscapes: classification, evaluation and conservation. *EEA, Environment Monographs 3.* European Environment Agency, Copenhagen, Denmark, 88 pp.

Gulinck, H. and Múgica, M. (1996) Landscape assessment at the European level. In: Wascher, D (ed.) *EEA European Landscapes: Classification, Evaluation and Conservation.* European Environment Agency, Copenhagen, Denmark.

Gulinck, H. and Múgica, M. (1997) Review of the EEA Report: European landscapes: classification, evaluation and conservation. European Environment Agency, Copenhagen, Denmark.

MANMF (1990) *Nature Policy Plan of the Netherlands.* Ministry of Agriculture, Nature Management and Fisheries, The Hague.

Magurran, A.E. (1988) *Ecological Diversity and Its Measurement.* Croom Helm, London.

Stanners, D. and Bourdeau, P. (eds) (1995) *Europe's Environment – The Dobríš Assessment.* A Report of the European Environment Agency, Copenhagen, Denmark.

Measuring the Impacts of Agriculture on Biodiversity

7

Graham Tucker

INTRODUCTION

It is widely acknowledged that modern agricultural practices have led to wide-spread losses and changes to biodiversity in Europe (e.g. Baldock, 1990; Tucker and Heath, 1994; Stanners and Bourdeau, 1995; Pain and Pienkowski, 1997). To overcome these and other environmental problems there is now a desire to develop policies, regulations and other initiatives that will ensure more sustainable agricultural systems. In order to assess the effectiveness of such initiatives it is essential to have reliable baseline and ongoing monitoring data on the state of the environment and the factors that impact on it. There is a considerable amount of information on the environment, much of which is published or stored in various databases. However, such a large quantity of information could swamp any analysis and make interpretation of impacts time-consuming, costly and difficult to interpret. Consequently, key messages could be missed. Furthermore, an even greater amount of data still needs to be collected on the state of biodiversity in Europe and the impacts of human activities thereon. Many species are poorly known and not surveyed or monitored at all, especially for taxa such as the lower plants, fungi, invertebrates, fish, amphibians, and even some mammals such as bats. To monitor and assess impacts on all species would clearly not be practical nor indeed desirable as it would massively exacerbate the problems of information overload described above.

To solve this problem, summary statistics are required that will identify and quantify the main issues for scientists, policy-makers and the general public. An increasingly popular method of achieving this is through the use of indicators. Indicators are quantified information on biotic or abiotic features that reflect to some degree the state of a landscape or components of biodiversity. Such indicators aim to fulfil three basic functions:

- simplification;
- quantification; and
- communication.

The objective should be to produce a limited number of indicators, so that the main trends are highlighted. The challenge is to strike a balance – the number of indicators should be as small as possible so that the main messages are clear but at the same time the issues must not be oversimplified. The danger with selecting too many indicators is that monitoring becomes ill-defined and inefficient.

Indicators should ideally address three separate but linked levels following the Organisation for Economic Co-operation and Development (OECD, 1997) driving force–state–response (DSR) framework, i.e.:

- the human and economic activities that exert *driving forces*, such as economic incentives, policy initiatives and the resulting agricultural practices, that can change the environment;
- the environmental condition or *state* that prevails as a result of the driving forces; and
- the *response* by farmers, consumers, industry and government to perceived changes in the state of the environment.

The objective of this chapter is to review the potential problems, limitations and benefits of using selective indicators for monitoring agricultural impacts on biodiversity. The monitoring of biodiversity can relate to a variety of scales, i.e. from the global to the local. As the use of indicators becomes increasingly efficient at larger scales of assessment, consideration will focus on their use for evaluating impacts at the international (i.e. CAP) and national levels. General properties of good indicators will be examined and suitable indicators identified and discussed. Lastly, a brief review of future data requirements will be made.

PROBLEMS

The use of indicators for monitoring agricultural impacts on biodiversity is difficult as a result of the following issues:

- the numerous habitats and species involved and the complex interactions between them;
- our currently poor understanding of the relationships between human activities and species, habitats and ecosystem processes;
- uncertainty and current debate over short- and long-term biodiversity conservation objectives;

- the effects of scale on assessments;
- the incompatibility of data sets and the paucity of information on many species and habitats.

DRIVING-FORCE INDICATORS

The monitoring of driving forces, such as economic policies and farming practices (e.g. pesticide or fertilizer use), can provide useful early warnings of potential problems. Where such data are collected over periods of time, these can also be used to investigate possible causes of unexplained changes in habitats or species populations, through, for example, correlative analysis. However, broad driving-force indicators are likely to be poor predictors of change in biodiversity. For example, the core indicator for environmental pressures on biodiversity and landscape proposed by the OECD Group on the State of the Environment is 'habitat alteration and land conversion from natural state' (OECD, 1994). This is likely to be a very poor predictor of overall changes in biodiversity for several reasons. First, this indicator assumes that there is a simple relationship between the loss of natural habitats and biodiversity. However, in Europe conversion of natural habitats to semi-natural habitats can often lead to complex changes in biodiversity; for example, overall increases in biodiversity but the loss of certain specialist species. Also, the relationship depends on the scale at which biodiversity is measured and the degree of existing habitat and landscape heterogeneity.

Secondly, it is very difficult to define what a natural habitat is. For example, it is unclear to what extent steppic habitats in southern Europe are the result of human activities (Suárez *et al.*, 1997). Similarly, it is difficult to define what is a significant habitat alteration or conversion. Some changes may be a very subtle and gradual response to variation in a continuous rather than discrete variable. For example, increases in stocking rates may lead to significant changes in vegetation composition and structure which, in turn, may lead to changes in associated animal communities (Ausden and Treweek, 1995).

Lastly, the indicator is not appropriate for artificial agricultural habitats in Europe, such as arable crops. These habitats predominate over much of western Europe and many of the most important impacts on biodiversity are now occurring within them (Pain and Pienkowski, 1997; Tucker and Evans, 1997).

A solution to such limitations of driving-force indicators would be to identify a range of more specific key influences on biodiversity, such as pesticide use. However, it is inevitable that, because of the huge number of species and the complexity of ecological processes within agricultural habitats, many potentially influencing factors may be unrecognized and not monitored. Furthermore, impacts such as pesticide use are often poorly understood, so the most relevant parameters that can practically be monitored are unclear. For example, the quantity of active pesticide ingredient used is often recorded, but this is unlikely to be useful in predicting, or explaining, impacts on biodiversity, as impacts differ according to the chemical compounds used, their concentration, method and timing of application, and the species involved. Driving-force indicators therefore cannot reliably predict actual changes in biodiversity. Indicators of the actual state of biodiversity are also essential.

INDICATORS OF THE STATE OF BIODIVERSITY

Habitat Extent

Perhaps one of the simplest indicators of the state of biodiversity is the extent of habitats. Such indicators can be linked readily to biodiversity conservation targets, with particularly important habitats monitored regularly and comprehensively, e.g. by satellite-based remote sensing and/or by sample surveys (Barr *et al.*, 1993; Tudor *et al.*, 1994). For example, in the UK the Biodiversity Steering Group (Anon., 1996) has identified 38 key habitats which are important because the UK has international obligations to protect them, they are rare or at risk, or they hold important species. A monitoring programme is being established under the Biodiversity Action Plan to measure changes in the extent of these key habitats. However, together these habitats only account for about 5% of the UK land area. Data on other habitats in the UK vary considerably. Comprehensive and detailed cropping data are available from artificial agricultural habitats but the extent of semi-natural habitats is much more poorly known (Gilbert and Gibbons, 1996).

At wider international scales, detailed habitat data are often lacking or at inappropriate levels of detail for use as biodiversity indicators (e.g. CORINE (co-ordination of information on the environment) land cover). Also, analyses across countries are often hampered severely by data incompatibility, especially relating to definitions and subdivisions of habitats. To overcome such problems the European Commission Concerted Action Project (CLAUDE) has been launched to link with other international programmes on this issue and to develop an internally consistent Europe-wide plan for land-use and land-cover monitoring and research.

Furthermore, and perhaps more importantly, data on the extent of a habitat do not give any information on its quality. For example, although the extent of a habitat such as wood pasture may be stable within an area, it may nevertheless be threatened by factors such as excessive grazing by livestock which results in little regeneration of trees. Also, information on the extent of a habitat does not offer any indication of the presence or absence of important species dependent on that habitat. In the UK, for example, many farmland birds have shown substantial declines despite relatively small changes in the extent of their arable and grassland habitats over recent decades (Fuller *et al.*, 1995). To solve these limitations biodiversity indicators must also take into account habitat quality and the presence of important species.

Habitat Quality

One method of assessing the general quality of a habitat is through the setting of common standards or quality measures. Such an approach is currently being developed in the UK for monitoring the condition of Special Areas of Conservation (SACs) and Sites of Special Scientific Interest (SSSIs) by the UK Statutory Conservation Agencies. Under this, generic guidelines are being prepared for the assessment of 'favourable condition' of habitats (Hill *et al.*, 1996a; Tucker *et al.*, 1997) according to the EU Council Directive on the Conservation of Natural Habitats and Wild Flora and Fauna (92/43/EEC). As an example of this approach, Humphries Rowell Associates (1995) suggest that the following attributes of a calcareous mesotrophic grassland may define 'favourable condition':

- open sward structure;
- no significant leaf litter;
- no poaching;
- rich in forbs including *Centaurea nigra*;
- presence of indicator species, e.g. *Ophioglossum vulgatum* and *Orchis morio*. Although these will vary with geographical area, this can often be accounted for in the preparation of guidelines by the provision of a list of suitable species, a number of which should be present.

As recommended by Rowell (1993), it is necessary to take account of deviations in a feature that are merely likely to be the results of natural variation or cyclic change. 'Limits of acceptable change' (LACs) should therefore be defined. If these threshold limits are then exceeded, the feature is no longer considered to be in 'favourable condition'. This approach enables the standardized and fairly simple recording of habitat quality where monitoring resources are limited. Such sample-based information on condition can be combined with habitat extent to give a more useful and reliable indicator of their state.

Species as Indicators of Biodiversity

As suggested in the above example of generic guidelines, the presence of particular indicator species can be used to assess the 'quality' of habitats and their overall biodiversity importance. However, this approach has its limitations, since to be good indicators for monitoring widescale impacts of land use on habitats and biodiversity species must be:

- specific to or highly concentrated in the habitat;
- widespread and relatively common in the habitat;
- easy to identify and have a well-established classification and systematics;
- easy to observe and census;

- well understood with respect to their ecology and interactions with agricultural land uses;
- closely linked to land-use practices and show rapid responses to these;
- representative of the habitat requirements and ecology of a large number of species;
- well monitored, with reliable baseline population estimates and long-term data (e.g. > 20 years) available at local, national and international scales;
- popular with the public to help motivate action (i.e. 'flagship' species).

On the basis of these factors a subjective assessment of the potential for different taxa groups to be used as indicators for monitoring impacts of agriculture on biodiversity are presented in Table 7.1. From this analysis, bird population data probably provide the best current potential for indicating the impacts of agriculture on biodiversity. Birds are relatively simple to identify and observe, have a well-established classification and systematics and are therefore comparatively easy to document with confidence. This, combined with their general popularity, has produced a wealth of information, much of which is collected by amateurs under the guidance of professional ornithological research bodies. Birds are also common and widespread, and because amateur work can cover a wide geographical area, broad-based monitoring of bird distributions, numbers and even reproductive success can be provided. Numerous ecological studies, such as those of habitat use, population regulation, feeding ecology and behaviour and migration have also been carried out by amateurs and professionals alike (e.g. Baille, 1990; Marchant *et al.*, 1990; Gibbons *et al.*, 1993; Tucker and Heath, 1994; Hagemeijer and Blair, 1997). As a result, bird populations and their general ecology are currently well understood and they are consequently often used as indicators of broad environmental changes (Peakall and Boyd, 1987; Furness *et al.*, 1993).

Birds may not, however, be the most sensitive of environmental indicators, as many species appear relatively resilient to change. Many bird species also appear to be relatively general in their habitat and food requirements and some species have adapted to non-intensive agricultural habitats where these have mimicked the broad structure of their original habitats. Although they tend to be high in the food chain and can therefore indicate disruption to food webs, responses may be slow and difficult to interpret. Bird numbers also tend to be regulated by density-dependent processes, so their population sizes may be buffered against environmental impacts (Furness *et al.*, 1993). Their migration movements may make it difficult to determine the original location where impacts may be occurring. On the other hand, the mobility of birds also facilitates the relocation of populations when conditions change or their recolonization if habitats recover.

Because birds may be less sensitive to environmental changes, declines in their populations arising from agricultural intensification may be less severe than in other taxa. Therefore, although we can assume that effects of agriculture on birds are likely to reflect effects on other taxa, as indicators they are

Table 7.1. Comparison of the potential for species of different taxa groups to act as indicators of impacts of agriculture on biodiversity.

Attribute	Mammals	Birds	Reptiles	Amph-ibia	Insects and Spiders	Other inver-tebrates	Higher plants	Lower plants
Many species concentrated in agricultural habitats	*	**	*	*	***	**	***	*
Widespread and common in agricultural habitats	**	***	*	**	***	***	**a	*
Easy to identify	***	***	**	**	*	*	**	*
Easy to observe and census	*	***	*	**	**	*	***	*
Well-understood ecology and interactions with agriculture	**	**	*	*	**	*	***	**
Sensitive to agricultural practices	**	**	*	*	***	**	***	**
Representative of a large number of other species	**	**	*	*	?	?	***	*
Well monitored at local, national and international scales	**	***	*	*	**	*	**	*
Potential as popular flagship species	***	***	**	*	*	*	**	*

a Originally much more common, but now scarce in many intensive agricultural habitats.
Suitability of the taxa with respect to the attribute: * = poor; ** = moderate; *** = good; ? = uncertain.

likely to underestimate overall impacts on biodiversity. Indeed, wild plant populations have already declined so much in Europe, many formerly widespread and common species are now completely absent from many agricultural habitats (e.g. Kornas, 1983; Hodgson, 1987; Van Dijk, 1991). Consequently, although sensitive to agricultural practices, their absence from many more intensively farmed habitats limits their capacity for use as indicators of further agricultural impacts. Plants and most of the other taxa groups listed in Table 7.1 are also of limited use as indicators due to a general lack of baseline population estimates and monitoring data.

Whichever species or species groups are chosen as indicators, it is essential that the relationship between the indicator and its related biodiversity objective is clearly understood. However, such relationships between indicators and their habitats and other species have rarely been demonstrated (Furness and Greenwood, 1993; Rowell, 1994; Stoltze and Wind, 1995). Therefore, the selection process must incorporate an objective test to prove that the selected indicator unambiguously reflects the specific performance targets for which it was chosen as an indicator. It is therefore recommended that a careful process is followed for the selection of performance indicators, as for example developed by Hunsaker and Carpenter (1990) for European Nature Indicators.

In some cases, the use of single-species indicators has been proposed as a means of monitoring impacts on biodiversity. However, this approach has severe limitations particularly when applied to broad habitat categories and on wide geographical scales. Individual species have different ecological requirements and therefore a change in the population of one may be due to a threat that is unique to it. Such information should not therefore be used to predict impacts on other species within the habitat, even where these have apparently similar ecological requirements.

To overcome such problems, data from a range of species are frequently combined to indicate impacts on biodiversity. Some of the apparently simplest of such indicators are species richness and diversity. However, the measurement of these is not straightforward (Magurran, 1988) and can be very time-consuming (as complete inventories, and counts in the case of diversity, are necessary). Also, richness measurements depend on the scale at which they are measured, invariably increasing with the size of the area (Schoener, 1976, 1986; Wiens, 1989). In agricultural habitats, such a species–area relationship is likely to occur as a result of environmental heterogeneity. Increasing the area will include additional habitat types, or variations within these, depending on how habitats are defined. Furthermore, the assessment of species richness is only one of several measures used to evaluate the importance of a habitat for biodiversity conservation. Its main limitations are that it does not take into account the conservation status of species or their reliance on the habitat in question. As mentioned above under driving-force indicators, degradation of some natural habitats can actually result in an increase in overall biodiversity even though specialized species of high conservation importance may be lost.

In order to address this, indicators have been developed that incorporate the conservation status of species. For example, in the UK, the proportion of four groups of species (fish, invertebrates, seed plants and mosses) that are considered 'threatened' according to criteria devised by the International Union for the Conservation of Nature (IUCN) or, where available, are 'nationally scarce' (i.e. recorded as present in 16–100 10-km squares in Great Britain) is proposed as an indicator of sustainable development (DoE, 1996). Similarly, OECD (1994) proposed the use of the proportion of total species known that are threatened or extinct as a core indicator of the state of biodiversity.

These indicators, however, have certain drawbacks. First, they concentrate on threatened species as defined by IUCN (commonly referred to as species on the Red Data List). Ideally, it would be better to include all species that are showing significant declines before populations become sufficiently small for the species to be regarded as 'threatened'. For example, in the UK the previous Red Data List for birds (Batten *et al.*, 1990) has been expanded to a wider list of species of conservation concern and includes species that have declined by more than 25% over the preceding 25-year period (Gibbons *et al.*, 1996). However, this new approach has yet to be applied commonly throughout Europe or to other taxa.

A second problem with using the proportion of threatened species as an indicator is that it is difficult to combine and interpret data from different taxa or countries as their interpretations of the IUCN criteria differ slightly. The newly revised IUCN criteria of Mace and Stuart (1994) are more quantitative and less subjective, but these have yet to be widely applied and will also require adaptation for national assessments which will inevitably introduce differences in application.

However, these problems do not apply to a recent assessment of the conservation status of birds in Europe (Tucker and Heath 1994) which used quantitative criteria applied across Europe to identify species with an 'unfavourable conservation status' (i.e. species that are declining, rare or highly localized). Thus the proportion of birds with an 'unfavourable conservation status' may indicate the state of biodiversity at a European scale. Furthermore, as shown in Table 7.2, these data can be analysed by habitat to identify those that are under proportionally most pressure (as indicated by the percentage of species with an 'unfavourable conservation status'). Clearly, it would be desirable to widen this approach to other taxa to compare results and improve the reliability and generality of this indicator with respect to predicted impacts on other components of biodiversity.

RESPONSE INDICATORS

The core biodiversity response indicator proposed by OECD (1994) is the proportion of national territory designated as protected areas for each ecosystem type. Clearly this is a 'means' indicator and does not directly measure the

Table 7.2. The number of bird species regularly occurring in agricultural habitats in Europe according to their conservation status and proportional use of each habitat (source: Tucker, 1997).

Habitat	Total number of species	Number of species with majority of population in habitat	Number of species with an 'unfavourable conservation status' (UFC)	Number with majority in habitat and UFC
Arable and agriculturally improved grass	122	20 (16%)	52 (43%)	12 (10%)
Steppic	80	28 (35%)	53 (66%)	24 (30%)
Wet grasslands	64	3 (5%)	30 (47%)	2 (3%)
Moorland	27	0 (0%)	12 (44%)	0 (0%)
Alpine grasslands	42	12 (29%)	24 (57%)	5 (12%)
Orchards, olive groves and perennial crops	59	0 (0%)	23 (40%)	0 (0%)
Pastoral woodlands	56	2 (4%)	23 (41%)	2 (4%)
Rice cultivations	36	1 (3%)	20 (55%)	1 (3%)

Figures in brackets indicate the percentage of the total species in the habitat.

effectiveness of the response in terms of its 'ends objectives', i.e. the conservation of biodiversity. This is a particular problem with the use of protected areas as an indicator, as it is widely documented that, even when protected, a large proportion of sites are significantly damaged or even destroyed (Stanners and Bourdeau, 1995). Thus, this indicator is not a good predictor of the effectiveness of the response.

Furthermore, the OECD indicator only provides useful information on the implementation of one aspect of conservation policy. Vital though the protection of important sites is, it is now becoming widely accepted that more emphasis needs to be placed on conservation measures in the wider environment (Tucker and Heath, 1994; Council of Europe/UNEP/ECNC, 1996; Tucker and Evans, 1997). Many species are widely dispersed, at least at some point in their annual cycle, and therefore cannot be adequately protected at a few specific sites. Furthermore, protected areas are not independent of the wider environment and are therefore influenced by activities in their surroundings. For example, wetlands are profoundly influenced by activities in their entire catchment, while coastal habitats are ultimately dependent on long-term and wide-scale geomorphological processes. Lastly, even when protected, the conservation of habitats and species within a site is often dependent on appropriate management which is usually influenced primarily by land-use regulations and policies. For example, protected semi-natural agricultural habitats such as alpine pastures are currently suffering from overgrazing encouraged by

Common Agricultural Policy (CAP) livestock headage payments (Donázar *et al.*, 1997).

To overcome such problems, response indicators should include quantitative performance targets based directly on conservation ends objectives. For example, at the local level, the monitoring of the effectiveness of 'environmentally sensitive areas' (under Agri-environment Regulation 2078/92) in the UK is based on their stated environmental objectives and publicized performance indicators (MAFF, 1994). Such performance indicators for birds in the Pennine Dales Environmentally Sensitive Area (ESA) are:

- an upward trend (subject to national trends) in the breeding population of yellow wagtail (*Montacilla flava*) in hay meadows;
- no downward trend (subject to national trends) in the breeding populations of waders, in particular lapwing (*Vanellus vanellus*), curlew (*Numenius arquata*) and redshank *(Tringa totanus)* in rough grazings and pastures.

At the national scale, the development of biodiversity plans according to the Convention on Biological Diversity provides a potential framework for the development of performance indicators based on biodiversity conservation targets. To demonstrate the feasibility of setting such measurable conservation targets, a consortium of voluntary bodies in the UK put forward over 530 species targets and 16 habitat targets (Wynne *et al.*, 1994).

For example, targets for some species of wet grassland include:

- *Apium repens* (creeping marshwort) – establish taxonomic validity of the last extant population in Oxfordshire and maintain appropriate management conditions for it.
- *Scozonera humilis* (viper's grass) – maintain at all sites in Dorset.
- *Segmentina nitida* (a freshwater snail) – maintain existing range and restore, if possible, to former sites in Norfolk and Suffolk Broads, East Kent and East Sussex.
- *Aeshna isoscles* (Norfolk aeshna dragonfly) – maintain existing populations. Expand area of suitable grazing marsh ditches by 50%.
- *Limosa limosa* (black-tailed godwit) – maintain as a breeding species in the UK. In the long term, increase the breeding population to over 100 pairs. Establish at least one more centre of population, with more than ten breeding pairs in addition to the Ouse and Nene Washes.
- *Lutra lutra* (otter) – maintain present range in the UK and expand to a population level throughout the UK which will show signs of otter activity on 90% of all rivers.

Thus monitoring should move from simply reporting changes in biodiversity towards setting positive targets against which assessments of the effectiveness of responses are made (Hill *et al.*, 1996b). In particular the success of biodiversity conservation initiatives should be assessed by answering the following questions:

- Are conservation targets being met?
- Are proposed actions still the right ones to meet the desired targets?
- Have the priorities for action changed?

DATA REQUIREMENTS

In order to develop and implement the use of indicators for monitoring impacts of agriculture on biodiversity, further land-use, ecological and biodiversity data are required. These include the following:

- comprehensive, detailed and integrated audits and maps of habitat cover at national and European level;
- sample-based surveys of habitat quality according to standardized criteria for defining 'favourable condition';
- baseline population estimates and sample-based monitoring data for a broad range of species;
- detailed information on the impacts of agricultural practices on species.

The collation of such data is clearly dependent on increased funding and coordination. This could be facilitated through the development of national and European biodiversity monitoring strategies. Such an initiative is currently under way for birds through the preparation of a European bird monitoring strategy by the European Bird Census Council in association with BirdLife International. Such data could then be used to assess the status of species in all taxa groups, using consistent approaches at national and international levels (e.g. as carried out for Bird Species of Conservation Concern in the UK (Gibbons *et al.*, 1996) and Species of European Conservation Concern (Tucker and Heath, 1994)). This could then be used as a consistent basis for the comprehensive setting of biodiversity conservation targets.

CONCLUSIONS

Using the indicator approach for monitoring the impacts of agriculture on biodiversity has several problems and limitations. First, monitoring driving-force indicators is hampered by insufficient understanding of the impacts of agricultural and other land uses and practices on habitats and their species. Consequently, although the collation of land-use data is useful for investigating possible causes of change, such indirect indicators are unreliable in measuring or predicting real impacts. Furthermore, the large number of species of agricultural habitats means that a huge number of potentially influencing factors exist, only a small proportion of which could ever be monitored practically.

Monitoring indicators of the state of biodiversity is also extremely difficult, again due to the potentially large number of species involved. Furthermore, as individual species have, to varying degrees, different habitat requirements and respond in differing ways to human impacts, it is not possible to select single, or even a small number of, species that can indicate the state of biodiversity in general. Monitoring of a broad range of species is therefore required. However, for many taxa groups, baseline or ongoing monitoring of data are not available.

Indicators of responses to biodiversity changes may be useful in assessing implementation of initiatives but are ultimately not reliable indicators of success in terms of ends objectives. Their success must be judged ultimately according to the state of biodiversity. It is suggested that the assessment of responses to impacts of agriculture on biodiversity should concentrate on the monitoring of their performance with respect to internationally, nationally and locally set quantitative biodiversity conservation targets.

REFERENCES

Anon. (1996) *Biodiversity: The UK Steering Group Report.* HMSO, London.

Ausden, M. and Treweek, J. (1995) Grasslands. In: Sutherland, W.J. and Hill, D.A. (eds) *Managing Habitats for Conservation.* Cambridge University Press, Cambridge, pp. 197–229.

Baille, S.R. (1990) Integrated population monitoring of breeding birds in Britain and Ireland. *IBIS* 132, 151–166.

Baldock, D. (1990) *Agriculture and Habitat Loss in Europe.* CAP Discussion Paper 3. WWF, Gland, Switzerland.

Barr, C.J., Bunce, R.G.H., Clarke, R.T., Fuller, R.M., Furse, M.T., Gillespie, M.K., Groom, G.B., Hallam, C.J., Hornung, M., Howard, D.C. and Nexx, M.J. (1993) *Countryside Survey 1990 Main Report.* Department of the Environment, HMSO Eastcote.

Batten, L.A., Bibby, C.J., Clement, P., Elliot, G.D. and Porter, R.F. (1990) *Red Data birds in Britain.* T & AD Poyser Ltd, London.

Council of Europe/UNEP/ECNC (1996) *The Pan-European Biological and Landscape Diversity Strategy.* Council of Europe/UNEP/ECNC.

DoE (1996) *Indicators of Sustainable Development for the United Kingdom. Consultation Paper.* Department of the Environment, London.

Donázar, J.A., Naveso, M.A., Tella, J.L. and Campión, D. (1997) Extensive grazing and raptors in Spain. In: Pain, D.J. and Pienkowski, M.W. (eds) *Farming and Birds in Europe.* Academic Press, London, pp. 117–149.

Fuller, R.J., Gregory, R.D., Gibbons, D.W., Marchant, J.H., Wilson, J.D., Baillie, S.R. and Carter, N. (1995) Population declines and range contractions among farmland birds in Britain. *Conservation Biology* 9, 1425–1442.

Furness, R.W. and Greenwood, J.J.D. (eds) (1993) *Birds as Monitors of Environmental Change.* Chapman & Hall, London.

Furness, R.W., Greenwood, J.J.D. and Jarvis, P.J. (1993) Can birds be used to monitor the environment? In: Furness, R.W. and Greenwood, J.J.D. (eds) *Birds as Monitors of Environmental Change.* Chapman & Hall, London, pp. 1–41.

Gibbons, D.W., Reid, J.B. and Chapman, R.A. (1993) *The New Atlas of Breeding Birds in Britain and Ireland: 1988–1991.* T & AD Poyser Ltd, London.

Gibbons, D.W., Avery, M.I., Baillie, S.R., Gregory, R.D., Kirby, J., Porter, R.F., Tucker, G.M. and Williams, G. (1996) Bird species of conservation concern in the United Kingdom, Channel Islands and the Isle of Man: revising the Red Data List. *RSPB Conservation Review* 10.

Gilbert, G. and Gibbons, D.W. (1996) A Review of Habitat Land Cover and Land-Use Survey and Monitoring in the United Kingdom. Royal Society for the Protection of Birds, UK.

Hagemeijer, W.J.M. and Blair, M.J. (1997). *The EBCC Atlas of European Breeding Birds: Their Distribution and Abundance.* T & AD Poyser Ltd, London.

Hill, D., Peterken, G., Rich, T. and Tucker, G. (1996a) *Generic Guidelines for HSD & Birds Directive (Contract No. F71-12-408).* Report to JNCC. Ecoscope Applied Ecologists, Cambridge.

Hill, D.A., Treweek, J., Yates, T. and Pienkowski, M. (1996b) *Actions for Biodiversity in the UK: Approaches in UK to Implementing the Convention on Biological Diversity.* Ecological Issue No. 6., British Ecological Society.

Hodgson, J.G. (1987) Growing rare in Britain. *New Scientist* 1547, 38–39.

Humphries Rowell Associates (1995) Common standards monitoring of SSSIs in practice. Report to JNCC, Peterborough.

Hunsaker, C.T. and Carpenter, D.E. (eds) (1990) *Ecological Indicators for the Environmental Assessment Program. EPA 600/3-90/060.* US Environmental Protection Agency, New York.

Kornas, J. (1983) Man's impact on flora and vegetation in central Europe. *Geobotany* 5, 277–286.

Mace, G. and Stuart, S. (1994) Draft IUCN Red List categories. *Species* 21/22, 13–24.

MAFF (1994) *Environmentally Sensitive Areas Scheme Environmental Objectives and Performance Indicators.* MAFF, London.

Magurran, A.E. (1988) *Ecological Diversity and its Measurement.* Croom Helm, London.

Marchant, J.H., Hudson, R., Carter, S.P. and Whittington, P. (1990) *Populations Trends in British Breeding Birds.* British Trust for Ornithology, Tring.

OECD (1994) *Environmental Indicators: OECD Core Set.* OECD, Paris.

OECD (1997) *Environmental Indicators for Agriculture.* OECD, Paris.

Pain, D.J. and Pienkowski, M.J. (eds) (1997) *Farming and Birds in Europe: the Common Agricultural Policy and its Implications for Bird Conservation.* Academic Press, London.

Peakall, D.B. and Boyd, H. (1987) Birds as bio-indicators of environmental conditions. In Diamond, A.W. and Filion, F.L. (eds) *The Value of Birds.* ICBP, Cambridge (Technical Publication 6), 113–118.

Rowell, T.A. (1993) *Common Standards for Monitoring SSSIs.* Report to JNCC, Peterborough.

Rowell, T.A. (1994) *Ecological Indicators for Nature Conservation Monitoring.* Joint Nature Conservation Committee Report No. 196. JNCC, Peterborough.

Schoener, T.W. (1976) The species–area relation within archipelagos: models and evidence from island land birds. In: Frith, H.J. and Calaby, J.H. (eds) *Proceedings 16th International Ornithological Conference.* Australian Academy of Sciences, Canberra, pp. 629–642.

Schoener, T.W. (1986) Patterns in terrestrial vertebrate versus arthropod communities: do systematic differences in regularity exist? In: Diamond, J. and Case, T.J. (eds) *Community Ecology.* Harper & Row, New York, pp. 556–586.

Stanners, D. and Bourdeau, P. (eds) (1995) *Europe's Environment – The Dobríš Assessment*. European Environment Agency, Copenhagen.

Stoltze, M. and Wind, P. (1995) *Nature Indicators Survey*. Report to the European Topic Centre on Nature Conservation, Paris. Ministry of Environment and Energy, National Environmental Research Institute. Draft Report.

Suárez, F., Naveso, M.A. and de Juana, E. (1997) Farming and the drylands of Spain: birds of the pseudosteppes. In: Pain, D.J. and Pienkowski, M.W. (eds) *Farming and Birds in Europe: the Common Agricultural Policy and its Implications for Bird Conservation*. Academic Press, London, pp. 297–330.

Tucker, G.M. (1997) Priorities for bird conservation in Europe: the importance of the farmed landscape. In: Pain, D.J. and Pienkowski, M.W. (eds) *Farming and Birds in Europe: the Common Agricultural Policy and its Implications for Bird Conservation*. Academic Press, London, pp. 79–116.

Tucker G.M. and Evans, M.E. (1997) *Habitats for Birds in Europe: A Strategy for their Conservation in the Wider Environment*. BirdLife Conservation Series No. 6. BirdLife International, Cambridge.

Tucker, G.M. and Heath, M.F. (1994) *Birds in Europe: Their Conservation Status*. BirdLife Conservation Series No. 3. BirdLife International, Cambridge.

Tucker, G.M., Hill, D., Peterken, G., Rich, T. and Elliot, M. (1997) *Generic Guidelines for HSD & Birds Directive (Contract No. F71-12-408)*. Report to JNCC. Ecoscope Applied Ecologists, Cambridge.

Tudor, G.J., Mackey, E.C. and Underwood, F.M. (1994) *The National Countryside Monitoring Scheme: the changing face of Scotland 1940s to 1970s*. Technical Report. Perth, Scottish Natural Heritage.

Van Dijk, G. (1991) The status of semi-natural grasslands in Europe. In: Goriup, P.D., Batten, L.A. and Norton, J.A. (eds) *The Conservation of Lowland Dry Grassland Birds in Europe*. Proceedings of an international seminar held at the University of Reading 20–22 March 1991. JNCC, Peterborough, pp. 15–36.

Wiens, J.A. (1989) *The Ecology of Bird Communities: Foundations and Patterns*, Vol. 1. Cambridge University Press, Cambridge.

Wynne, G., Avery, M., Campbell, L., Juniper, T., King, M., Smart, J., Steel, C., Stones, T., Stubbs, A., Taylor, J., Tydemen, C. and Wynde, R. (1994) *Biodiversity Challenge*. Royal Society for the Protection of Birds, Sandy, Bedfordshire.

Nature Quality Indicators in Agriculture

8

Jaap van Wenum, Jan Buys and Ada Wossink

INTRODUCTION

The European Commission's Fifth Environmental Action Programme *(Towards Sustainability)* and the Reform of the Common Agricultural Policy in 1992 have both initiated new legal and strategic actions to integrate environmental objectives into sectoral policies. Agriculture, in particular, is considered a driving force affecting the state of, and changes in, Europe's biological and landscape diversity. The former natural vegetation of the continent of Europe has been changed virtually in its entirety under the influence of human land-use activities; large proportions of existing and future ecological values are more or less directly dependent on the way the land is utilized.

In The Netherlands the issue became a topic of public and political interest in the 1980s. It has led to new policies on land use, agricultural practices, nature and landscape, namely the National Environmental Policy Plan (NEPP) of 1989 and the Nature Policy Plan (NPP) of 1990. The main focus of the NPP is the Ecological Main Structure (EMS), an ecological network consisting of core areas (nature reserves) and nature development areas, connected by means of corridors. Outside the EMS, where agriculture is the dominant land use, NPP envisages a minimum environmental quality, supplied by farmers and other land users. However, the implementation of this part of the NPP policy plan is still in its infancy. With regard to nature conservation on farms, the focus is now on the development and introduction of new incentive schemes based on nature, supplied, in addition to the current practice, with command and control schemes (Van Paassen *et al.*, 1991; Van Harmelen *et al.*, 1995; MLNV, 1995; Ter Steege *et al.*, 1996). For the NPP implementation, both regarding the EMS and the incentive schemes, the need is felt for indicators showing 'nature quality' and its development over time (see also Udo de Haes *et al.*, 1993; Nature Conservation Council, 1995). Indicators are crucial instruments as they enable goal setting and impact evaluation. Moreover, they can serve as a basis for conservation payments to farmers, and provide a basis for certification schemes for agricultural products (Buys, 1995; Horlings and Buys, 1997).

For farmers, indicators can be a useful guide to management decisions directed towards nature conservation by providing information on the ecological benefits of modifications to farming practice and of specific nature conservation measures. This kind of information is also useful for economic research on agriculture and nature conservation. Several economic tools are available which incorporate the relationships between agricultural output, environmental quality indicators and production techniques focusing at the crop or whole-farm level. Partial budgeting and programming methods are the predominant methods (for an overview see Roberts and Swinton, 1996). Typically, these methods are used to gain insights into the trade-offs between income and environmental stress (see, for example, Wossink *et al.*, 1992; Foltz *et al.*, 1995; Teague *et al.*, 1995; Verhoeven *et al.*, 1995). These approaches can also be used to study the implications of nature conservation at the farm level. With a nature quality indicator, account can be taken of both the economic effects of nature conservation measures and the ecological consequences. Constrained optimization procedures may be used to define best management strategies at farm level for each targeted nature quality index. Finally, generally accepted quality indicators facilitate communication between the parties involved.

This chapter focuses on nature quality indicators and the specific requirements involved when they are applied to agriculture. Special attention is drawn to the so-called *yardstick for biodiversity*, an indicator developed recently by the Centre for Agriculture and Environment (Buys, 1995).

In the second section, some conceptual considerations on nature and biodiversity and on criteria and indicators for nature quality are presented, and this is followed by a review of recent developments in designing indicators and on specific requirements for nature quality indicators when applied in agriculture. Then the recently developed yardstick for biodiversity is presented and initial experience with its use analysed. The chapter concludes with a discussion and outlook.

CONCEPTUAL CONSIDERATIONS

Nature and Biodiversity

Nature is often defined as everything that organizes and sustains itself, whether associated with human action or not (see also Veeneklaas *et al.*, 1994; Nature Conservation Council, 1995). This definition implies different gradations ranging from almost natural (e.g. jungle) to cultural (e.g. roadsides). The value of nature is often associated with the biotic (living) part of the natural environment. The term 'biological diversity', usually shortened to biodiversity, is commonly used to indicate the total biotic environment. Biodiversity is an umbrella term for the number, variety and variability of the living organisms in a given assemblage (Pearce and Moran, 1994). So the assemblage, or abiotic natural environment, is considered conditional for the state of the biodiversity. Hence, biodiversity is a useful representation of the quality of the natural environment.

Biodiversity may be described in terms of genes, species and ecosystems, corresponding to three fundamental and hierarchically related levels of biological organization (Pearce and Moran, 1994). First, genetic diversity is the sum of genetic information contained in the genes of individual plants, animals and microorganisms; secondly, species diversity considers the richness, variation or number of different species, and populations within which gene flow occurs under natural conditions; and, thirdly, ecosystem diversity relates to the variety of habitats, biotic communities and ecological processes in the biosphere as well as the diversity within ecosystems.

Valuing Nature

Veeneklaas *et al.* (1994) summarize the two principal views that exist with respect to the value of nature:

1. Ecocentric: nature has an intrinsic value independent of its utility to mankind, and conservation of each individual species is therefore equally relevant.
2. Anthropocentric: nature has a value to humans as nature fulfils different functions:

- production function, including production of renewable and non-renewable resources;
- carrier function: nature as space provider for human activities;
- information function: the availability of nature for recreation, science, education;
- regulation function: stabilizing functions of nature to environmental changes such as (water) purification capacity and resistance and retention features of nature.

In the environmental economics literature it is common to consider the total economic value of nature as the aggregate of all user- and non-user-motivated values. Although the user and existence point of view might be complementary, the associated values might be overlapping (Holstein, 1996). Considering the values, i.e. the functions nature performs, its interrelationships with agricultural land use are obvious. Hence it is no surprise that the main causes for the deterioration of the natural environment are related directly to agricultural practice (Terwan and Van der Bijl, 1996): loss and fragmentation, acidification, eutrophication and desiccation.

Nature Quality Criteria and Indicators

Although the functions performed by nature can be defined, these are difficult to address in policy design. The NPP (1990) concentrated specifically on biodiversity. Three criteria of biodiversity were specified as objectives in policy development: (i) naturalness, (ii) diversity (in terms of (inter)national rarity of species) and (iii) 'characteristic features'. 'Naturalness' not only includes the degree of human intervention but also the size and completeness of ecosystems. 'Characteristic features' seems a redundant criterion as one would expect that under more or less natural conditions (i.e. no human interference) only these species and ecosystems will be present that are characteristic of these conditions (Bal *et al.*, 1995). Direct measurement using these criteria is difficult and there is a role for indicators of performance that may directly or more tangentially relate to the criteria specified.

In this chapter the focus is on indicators for agricultural areas, where nature coexists with human activities (agriculture). Therefore naturalness is not a relevant criterion. Hence, in context of the criteria outlined above, nature quality indicators for agricultural areas are defined only in the context of the diversity criterion.

The Organisation for Economic Co-operation and Development (OECD) work on agriculture and the environment recognized pressure, state and response (PSR) indicators (OECD, 1994). Hence, the PSR approach is applied to (bio)diversity in agriculture. Pressure indicators are measurements of agricultural activities that cause changes in the state of biodiversity, such as the use of pesticides and fertilizer. State indicators are direct measurements of the state of biodiversity arising from these pressures, in terms of species, habitats or environmental parameters. Finally, response indicators refer to responses by farmers, government or society to changes in the state of biodiversity, such as the use of financial incentives to farmers to enhance biodiversity. According to Udo de Haes *et al.* (1997), pressure indicators are less relevant as the concept of nature quality is effect-orientated, and therefore indicators of impacts in terms of species or environmental parameters are preferable. Moreover, many of the relationships between human activities and the abundance of species and habitats are complex and poorly understood. With regard to state indicators, two types can be distinguished: (i) indicators that are direct measurements of species biodiversity; and (ii) indicators that are measurements of environmental quality or biotope (habitat) presence, conditional for the presence and abundance of wildlife. Examples of both types of state indicators are discussed in the next section. Response indicators go beyond the aims of this chapter and will not be discussed.

DESIGNING NATURE QUALITY INDICATORS

Some Reflections on Recently Developed Indicators

Recently developed indicators for nature quality have been based mainly on the diversity concept, and indicators considering both species diversity and ecosystem (biotope or habitat) diversity have been developed.

An example of a biotope-diversity-based nature quality indicator for agricultural areas is described by Wahlberg-Jacobs (1991). Arable areas in Germany are assessed in terms of the density and size of landscape elements (hedges, ditches, tree groups, etc.) as well as agricultural practices such as pesticide use, nitrogen utilization and farming system (conventional, organic). The method also takes into account soil type and susceptibility to erosion. The resulting value indicates the nature tolerance of the studied area. Smeding (1995), who explored the nature quality characteristics of organic farms, also measured the area and variety of biotopes. Although such methods consider the quantitative and qualitative biotope requirements for the presence of wildlife, information on the actual presence of species is not recorded. Species-diversity-based indicators usually depart from the true presence and abundance of species and give more direct information on nature quality.

Many attempts have been made to measure biodiversity by constructing species diversity indices. Ecologists often construct biodiversity indices as a function of species counts and the relative abundance of species (Magurran, 1988). Species richness is the simplest form of index, but it omits differences in abundance between species in a set. Economists have taken a different approach by constructing biodiversity indices as a function of genetic distances among members of a species set (Solow *et al.*, 1993). Species richness fails to incorporate this concept.

An example of a species-diversity-based indicator that has also been used to define policy objectives is the AMOEBA approach of Ten Brink (1991). In order to examine the quality of the North Sea ecosystem and its development over time, as well as to define targets, it combines a reference population size (in the year 1930), current population size and target size of a representative selection of marine species. The focus is on reaching the reference size in a predetermined period. Target sizes of species reflect policy objectives to move the system towards this situation. Criticism of this approach has focused on the representativeness of the species set, the unpredictability of nature and corresponding interpretation of the quantitative goals (De Bruin *et al.*, 1992; Van der Windt, 1996). More recently, the method has been applied to other (aquatic and terrestrial) ecosystems.

Species diversity indicators linked to nature protection in the nature policy of The Netherlands have also been developed. This approach determines the protection need of species in terms of international importance, rarity and population development (tendency). An example is a vascular plant species-based indicator for vegetation developed by the province of South-Holland

(Clausman *et al.*, 1984). Species values were based on current rarity at different scales (provincial, national and international) as well as population size development, and used, together with values for species cover, to calculate the floristic value of vegetation.

A more extensive method for valuing nature in agriculture, using the protection approach, has been developed recently by the Centre for Agriculture and Environment (CLM) (Buys, 1995). This yardstick for biodiversity not only involves vascular plants but includes mammals, birds, butterflies, amphibians and reptiles. Species values based on rarity, population development and international importance are used in conjunction with census methods to calculate the nature value of farms.

Sijtsma *et al.* (1995), when predicting the impact of the implementation of the EMS, used a similar methodology but different species selection since non-agricultural ecosystems were involved. Census methods were not used. Criticism of the methodology focused on the representativeness of the species set which consisted of so-called goal species – principally endangered species (Buys, 1996).

Nature quality dependent payment systems for use in agriculture have recently been developed and implemented by Leyden University (Mibi) and the Ministry of Agriculture, Nature Management and Fisheries (LBL) (Van Paassen *et al.*, 1991; Van Harmelen et al. 1995; Ter Steege *et al.*, 1996). For meadow birds, a system is used with species-specific payments, based on the protection need of the species. For field margin schemes a system is used based on the presence of a selected number of plant species. Payment levels under both schemes are indicative of the nature quality on the farms.

Demands for Agriculture

Within the agricultural sector different users have different requirements for nature quality indicators. Farmers require clear benefits from the assessment of nature quality. Nature quality indicators can be introduced successfully only when they are used within incentive or certification schemes. Since unlike other agri-environmental indicators (e.g. mineral accounts), they cannot be extracted from farm accounting data, and therefore require an additional labour input, it is important that assessment for nature quality can be performed by farmers and, from a policy perspective, indicators have greater value if they can be used in incentive schemes. Furthermore, it is important that information on nature quality is made available at farm scale. There is a deficiency of farm-level data as environmental information is normally gathered at a higher level (ecosystem, region). The farmer, however, bases decisions on available farming techniques and on farm economic considerations. Hence the farm level is also the level of aggregation at which the economic and the ecological interactions are most pronounced.

Taking the foregoing into account, the CLM yardstick and the Mibi/LBL methods appear most appropriate for application in agriculture. The Mibi/LBL methods are simple but relatively weak in terms of biotopes and species coverage. The CLM yardstick is more complex, requiring more knowledge and time, but covers a much broader range of species and also takes the whole farm into account (Horlings and Buys, 1997). The remaining part of this chapter therefore places emphasis on this indicator.

YARDSTICK FOR BIODIVERSITY

Basic Principles

The yardstick for biodiversity is an instrument for quantifying and valuing nature (biodiversity) on farms. The yardstick is based on the following main principles: (i) biodiversity refers to organisms which (may) establish and sustain spontaneously and is assessed at the level of individual species; (ii) at farm level a quantitative assessment of biodiversity is made of each biotope, although in the testing process on farms the yardstick was simplified by not splitting the farm into separate biotopes – such an assessment is based on a representative selection of (groups of) species; (iii) the yardstick assigns a rating to selected species expressing their significance to society; and (iv) yardstick scores are made up for each species group and indicate the value of nature on a farm as follows:

$$Y(s) = \sum_{b=1}^{B} \sum_{j=1}^{J} cv \qquad\qquad [8.1]$$

where:
$Y(s)$ = yardstick score of species group s;
B = number of biotopes on farm;
J = number of selected species in species group s;
c = census units of species j
v = value to society (rating) of species j.

The structure of the yardstick is such that it enables farmers to identify how biodiversity may be increased. The yardstick for biodiversity is applicable to all farm types (Buys, 1995).

Species Selection

The yardstick for biodiversity comprises a limited selection of species groups which:

- enables reliable and simple census methods to be used (that can be undertaken by farmers);
- provides information on the effect of farm management on biodiversity;
- reflects biodiversity on farms.

Based on these principles, vascular plants, larger mammals, birds, amphibians, reptiles and butterflies were selected. (Butterflies were selected for their relative importance to nature conservation on farms despite the lack of a simple quantitative census method for the group.) From each of these groups individual species were selected, taking into account the following criteria:

- presence of the species in agricultural areas;
- likelihood of encountering the species;
- correlation between the species and farm management;
- indicative value of the species for the condition of the biotope;
- recognition of the species.

These criteria resulted in a list of 199 species of vascular plants, 17 of mammals, 77 of nesting birds, 14 of wintering birds, 7 of amphibians, 2 of reptiles and 6 species of butterflies, with the aim of covering the entire country and all farmland habitats (Buys, 1995).

Census Methods

An essential aspect of the yardstick for biodiversity is a quantitative assessment of biodiversity for different biotopes on farms (Table 8.1). Quantitative census methods must be reliable, feasible for use by farmers and preferably comparable with other census projects (Buys, 1995).

Commonly used census methods for vascular plants are the methods of Braun-Blanquet and Tansley (Den Held and Den Held, 1992), but these methods require specific knowledge in order that species cover can be assessed. Recently, an alternative census method has been developed by the Central Statistical Bureau (CBS) which is supposedly easier to use. Both abundance, using nine number classes, and cover of species, using percentages, are estimated. For the yardstick for biodiversity, assessing plant cover is less relevant and therefore only the abundance is estimated using this CBS classification (Buys, 1995). During the process of testing on farms this classification was simplified and reduced to four number classes (1–10, 11–100, 101–1000, > 1000 plants).

Census methods available for mammals generally depart from counting individuals or counting the number of roosts or dens (De Wijs, 1994). The yardstick for biodiversity uses both methods according to species characteristics (Buys, 1995).

Four census methods for nesting birds are available (Hustings *et al.*, 1985): transect counts, the tally method, territory mapping and fledgling counts. The

Table 8.1. Census methods used by the yardstick for biodiversity.

Species group	Census method
Vascular plants	CBS classification
Mammals	Roost, den or number counts[a]
Nesting birds	Tally method and fledgling counts
Wintering birds	Abundance classification × length of stay
Butterflies	–
Amphibians	Egg batch/string counts or recording calling males[a]
Reptiles	Number counts

[a]In accordance with species characteristics.
CBS = Central Statistical Bureau.

first three methods are based on counting occupied territories and differ in the area to be covered by the observations. The yardstick for biodiversity uses the tally method as providing sufficient quantitative information with relatively limited effort. Fledgling counts are used alongside this method to assess the breeding success of species nesting on the productive parts of the farm. The advantage of the latter method is that the results illustrate clearly the relationship with farming practice. For wintering birds, transect counts, and counting individuals or faeces are available methods. The first method provides only limited quantitative information, and the last two are relatively time-consuming; none of the methods takes into account the length of stay of the wintering birds. The yardstick therefore uses an adapted census method which includes counting individuals to assess the maximum number present during winter (in four classes) and determining the length of stay in months (Buys, 1995).

As mentioned earlier, there is no relatively easy-to-use quantitative census method for butterflies: the yardstick therefore only assesses the presence of a butterfly species (Buys, 1995).

Recommended census methods for amphibians are counting egg batches or strings and counting individuals (calling males, Stumpel and Siepel, 1993). The yardstick uses both methods (using five number classes) in accordance with species characteristics (Buys, 1995). Reptiles are usually monitored by counting individuals (Stumpel and Siepel, 1993) and the yardstick also uses this method.

Rating of Species

Yardstick scores are made up by multiplying the number of units resulting from the farm census and a species rating score. This rating score expresses the importance society attaches to a species. The importance of a species to society depends on rarity, the degree to which it is endangered and attractiveness (NPP, 1990). The first two aspects relate to the ecological importance of species, the latter one relates to the contribution of a species to the scenery. The yardstick uses a rating system based on the ecological importance of species and this is used here, although for vascular plants a system based on the scenic value is also available (Buys, 1995).

In nature policy three aspects are important when considering the ecological importance of species (Clausman *et al.*, 1984; Bink *et al.*, 1994):

- rarity: population size (assessed regionally, nationally or internationally);
- tendency: development in population size (assessed regionally, nationally or internationally);
- international importance: the importance of the presence of a species in The Netherlands to the global survival of that species.

All three aspects were incorporated in the yardstick rating system and applied at national level. Rarity of species (R_j) was calculated by dividing the total number of topographical grid cells in The Netherlands (1677, each grid cell at 25 km^2) by the number of grid cells in which a species is found:

$$R_j = 1677 \text{ / number of topographical grid cells with species j found.}[1] \quad [8.2]$$

Tendency of species (T_j) was calculated by assessing the change in (national) population size in terms of percentage:

$$T_j = \left(\frac{(\text{past population size}_j - \text{current population size}_j)}{\text{past population size}_j} \right) \times 100. \quad [8.3]$$

Different time periods were considered when assessing population changes of different species groups, mainly due to limitations in availability of data. For species with constant or increased population sizes, T_j equals 0.

For different species groups, different criteria are used to assess international importance (Bink *et al.*, 1994). The number of criteria range from 1 (mammals, amphibians and reptiles) to 4 (nesting birds). International importance of species (I_j) was incorporated in the yardstick rating by determining the relative number of criteria met by a species:

$$I_j = \frac{\text{number of criteria for international importance species j meets}}{\text{number of criteria for international importance of respective species}} \quad [8.4]$$

The ecological rating of a species (E_j) was defined according to the following procedure (Buys, 1995):

1. Minimum values of components of ecological importance were put at 1.

2. The maximum value for international importance was put at 1.5 because of the relatively minor importance of this component to nature conservation at a national level.

3. Components of ecological importance were multiplied.

4. To achieve a rating range from 1 to 100 as well as distinction between (general) species values a logarithmic transformation was carried out.

The ecological rating of a species (E_j) was calculated as follows:

$$E_j = 18.5 \times \log\{R_j \times T_j[1 + (I_j \times 0.5)]\} \qquad\qquad [8.5]$$

Example

Table 8.2 presents an example of a yardstick result form based on a fictitious 20 ha farm with two biotopes: grassland (18 ha) and ditches (2 ha). Census results are used together with the ecological rating of species to make up biotope scores for each of the species groups. Biotope scores are then added up to produce farm scores for the respective species groups. Biotope and farm scores per hectare can also be calculated. It is not possible to aggregate species group scores to a single nature quality score per farm or per hectare as both census methods and the rating method differ for the various species groups. However, when more yardstick results become available a reference level for each species group may be defined and relative scores may be introduced. This would enable comparison and aggregation of species group scores.

Practical Experience

The presented yardstick for biodiversity has only recently been developed and an extensive practical test is in progress. So far, the yardstick has been used to analyse the results of an experiment to enhance natural values on set-aside land (Buys *et al.*, 1996). In 1995 a small-scale test was carried out on four dairy farms (Buys and Ter Steege, 1996). In 1996 another eight dairy farms tested the yardstick, and in 1997 it was applied to 20–30 farms (both dairy and arable). Limitations of the yardstick revealed by the 1995 and 1996 tests are discussed below.

In the set-aside experiment, the yardstick was used by ecological researchers, so no information on its ease of use by farmers was obtained. However, one limitation of the yardstick did appear. It became apparent that calculating biotope scores per hectare was inaccurate for vascular plants where census classes were used rather than true numbers. This reflected the relationship between the composition of the census classes and the area of the trial fields. Differences in field size could produce differences in biotope scores per hectare despite the presence of similar census clauses. This was unexpected given the actual densities of species and the field size ratio (Buys *et al.*, 1996).

Table 8.2. Example of a yardstick form based on a fictitious farm with two biotopes (after Buys, 1995).

	Vascular plants	Nesting birds	Wintering birds	Mammals	Amphibia and reptiles	Butterflies
Biotopes						
Grassland	318	170	10	10	0	15
Ditches	309	95	0	6	81	0
Farm total	627	265	10	16	81	15
Average score ha^{-1}	31.4	13.3	0.5	0.8	4.1	0.8

The test carried out on the dairy farms had two main objectives: first, to analyse the feasibility of farmers applying the yardstick methods; and, second, to analyse whether the yardstick scores represent the nature present on farms. Both the farmer and a field biologist carried out the same assessments. To analyse the representativity of the selected species, all species within yardstick species groups were rated and counted. Total scores obtained were compared to scores obtained when using the selected species set only.

Farmers found it relatively easy to trace and identify species with the exception of vascular plants. Some 30% of the yardstick species from this group that were present were not traced. Identification of the vascular plant species, on the other hand, was found to be adequate. The results found by the field biologist and the farmer differed for censuses of species groups where classes were used as opposed to true numbers (vascular plants, amphibians). With respect to the representativity of the species selection, a similar pattern of scores across farms and biotopes was obtained from the total inventory and the yardstick species inventory. However, as only a limited number of farmers were involved and only one farming type was considered, these results can only be interpreted as indications of the yardstick's opportunities and limitations (Buys and Ter Steege, 1996).

DISCUSSION

In order to assess nature quality on farms the yardstick for biodiversity uses an approach of quantifying and rating a set of representative species. The rating is highly linked to nature policy in The Netherlands towards endangered species and the requirement for species protection. From a nature consumers' perspective, however, the aesthetic value of nature is very important. A rating of species which incorporated aesthetic features would therefore be more appropriate. Such a rating is available for plants (Buys, 1995) but is subjective as preferences among consumers may vary.

In order to minimize practical problems in its use by farmers, the yardstick uses a limited number of species (groups) and easy census methods where applicable. However, the vascular plants group still seems to pose difficulties for application of the yardstick. Possible modifications, such as limiting the census area and using other census methods, should therefore be considered. For ecological research purposes, i.e. the analysis of ecological effects of nature conservation measures, the accuracy of the yardstick can be increased by carrying out a census of all species within a group or even other species groups, provided information that is needed to calculate a rating for these species is available. Other census methods may also be applied.

Before applying the yardstick for policy purposes, and more particularly in incentive schemes, the methods have to be tested extensively. So far, incentive schemes based on nature output have used the presence of a limited number of vascular plant species or nest numbers of meadow birds and species-specific payments (Van Paassen *et al.*, 1991; Van Harmelen et al. 1995; Ter Steege *et al.*, 1996). The disadvantage of the yardstick compared to these methods is that it is time-consuming and results in higher costs for farmers (labour) and policy executors (inspection). The farmers in the tests described above (15 ha farms) needed a total of 20–30 hours to carry out the census and maintain records (Buys and Ter Steege, 1996). Horlings and Buys (1997) acknowledge this drawback but state that a simplified version, using fewer species groups or biotopes, could enhance the potential of the yardstick for use in incentive schemes while maintaining some of its comparative advantages over other methods. Furthermore, provided the benefits to the farmer balance the efforts required in applying the yardstick in practice, they see a number of other future policy applications for:

- management agreements: adding a nature-quality-dependent premium on top of the current effort-based payments;
- tax and financial advantages to farmers in exchange for achieving minimum yardstick scores;
- farm certification;
- establishing standards for a minimum or desired nature quality.

With these applications in mind the focus is now on further testing and fine-tuning of the yardstick approach on a range of farm types in different regions.

NOTE

[1] For nesting birds a different measure was used (because of data availability): true population/theoretical maximum population; the theoretical maximum population equals the current population of the most abundant bird species in The Netherlands (blackbird, 1 million pairs).

REFERENCES

Bal, D., Beije, H.M., Hoogeveen, Y.R., Jansen, S.R.J. and van Reest, P.J. (1995) *Handboek Natuurdoeltypen in Nederland*. IKC Natuurbeheer, Wageningen, The Netherlands.

Bink, R.J., Bal, D., van den Berk, V.M. and Draaijer, L.J. (1994) *Toestand van de Natuur 2*. IKC Natuur, Bos, Landschap en Fauna; Ministerie van Landbouw Natuurbeheer en Visserij, Wageningen, The Netherlands.

Buys, J.C. (1995) *Naar een Natuurmeetlat voor Landbouwbedrijven*. Centrum voor Landbouw en Milieu, Utrecht, The Netherlands.

Buys, J.C. (1996) Meten van natuurwaarden op landbouwbedrijven. In: Strinjker, D. and Sijtsma, F.J. (eds) *Meten van Natuur*. Stichting Ruimtelijke Economie, Groningen, The Netherlands.

Buys, J.C. and Ter Steege, M.W. (1996) *Naar een Natuurmeetlat voor Landbouwbedrijven II: de eerste Praktijktest op vier Melkveebedrijven*. Centrum voor Landbouw en Milieu, Utrecht, The Netherlands.

Buys, J.C., Oosterveld, E.B. and Ellenbroek, F.M. (1996) *Kansen voor Natuur bij Braaklegging II*. Centrum voor Landbouw en Milieu, Utrecht, The Netherlands.

Clausman, P.M.H.A., van Wijngaarden, W. and den Held, A.J. (1984) *Het Vegetatieonderzoek van de Provincie Zuid-Holland*. Vol 1: Verspreiding en ecologie van wilde planten in Zuid-Holland, Part A: Waarderingsparameters. Provinciale Planologische Dienst Zuid-Holland, The Hague, The Netherlands.

De Bruin, J., van Hees, B.W.M., Praat, P.J.A., Swart, J.A.A., van der Windt, H.J. and Winter, H.B. (1992) *De Amoebe en Onzekerheden*. Afdeling Biologie/Vakgroep Bestuursrecht en Bestuurskunde, Rijksuniversiteit Groningen, Groningen, The Netherlands.

Den Held, J.J. and den Held, A.J. (1992) *Beknopte Handleiding voor Vegetatiekundig Onderzoek*. Wetenschappelijke mededelingen No. 97. Koninklijke Nederlandse Natuurhistorische Vereniging, Utrecht, The Netherlands.

De Wijs, W.J.R. (1994) *Zoogdiermonitoring – een Studie naar de Haalbaarheid van een Meetnet Zoogdieren*. Vereniging voor Zoogdierkunde en Zoogdierbescherming, Utrecht; Vleermuiswerkgroep Nederland, Wageningen, The Netherlands.

Foltz, J.C., Lee, J.G., Martin, M.A. and Preckel, P.V. (1995) Assessment of alternative cropping systems. *American Journal of Agricultural Economics* 77, 408–420.

Holstein, F. (1996) The values of the agricultural landscape – a discussion on value-related terms in natural and social sciences and the implications for CVM and policy measures. In: *Proceedings of a Workshop on Landscape and Nature Conservation, 26–29 September, 1996*. Department of Farm Management, University of Hohenheim, Germany.

Horlings, L.G. and Buys, J.C. (1997) *Zicht op Natuurresultaat, de Natuurmeetlat als Beleidsinstrument*. Centrum voor Landbouw en Milieu, Utrecht, The Netherlands.

Hustings, M.F.H., Kwak, R.G.M., Opdam, P.F.M. and Reijnen, M.J.S.M. (1985) *Vogelinventarisatie – Achtergronden, Richtlijnen en Verslaglegging*. PUDOC, Wageningen, The Netherlands.

Magurran, A.E. (1988) *Ecological Diversity and its Measurement*. Croom Helm, London.

MLNV (1995) *Dynamiek en Vernieuwing – Prioriteitennota*. Ministerie van Landbouw, Natuurbeheer en Visserij, The Hague, The Netherlands.

Nature Conservation Council (1995) *Natuur Buiten Natuurgebieden – Actief Beleid voor Algemene Natuurkwaliteit: een Handreiking*. Raad voor het Natuurbeheer, Utrecht, The Netherlands.

NEPP (1989) *Nationaal Milieubeleidsplan*. Ministerie van Volkshuisvesting, Ruimtelijke Ordening en Milieu, The Hague, The Netherlands.

NPP (1990) *Natuurbeleidsplan – Regeringsbeslissing*. Ministerie van Landbouw, Natuurbeheer en Visserij, The Hague, The Netherlands.

OECD (1994) *The Use of Environmental Indicators for Agricultural Policy Analysis*. COM/AGR/CA/ENV/EPOC (94) 48. OECD, Paris.

Pearce, D. and Moran, D. (1994) *The Economic Value of Biodiversity*. IUCN – The World Conservation Union, Earthscan Publications, London.

Roberts, W.S. and Swinton, S.M. (1996) Economic methods for comparing alternative crop production systems: a review of the literature. *American Journal of Alternative Agriculture* 11, 10–17.

Segerson, K. (1988) Uncertainty and incentives for nonpoint pollution control. *Journal of Environmental Economics and Management* 15, 87–98.

Sijtsma, F.J., Strijker D. and Rotmensen, G.J. (1995) *Effect-analyse Ecologische Hoofdstructuur Vol II: Natuurwaarde*. Stichting Ruimtelijke Economie, Groningen, The Netherlands.

Smeding, F.W. (1995) *Protocol Natuurplan*. Vakgroep Ecologische Landbouw, Landbouwuniversiteit Wageningen, The Netherlands.

Solow, A., Polasky, S. and Broadus, J. (1993) On the measurement of biological diversity. *Journal of Environmental Economics and Management* 24, 60–68.

Stumpel, A.H.P. and Siepel, H. (1993) *Naar Meetnetten voor Reptielen en Amfibieën*. Instituut voor Bos-en Natuuronderzoek, Wageningen, The Netherlands.

Teague, M.L., Bernardo, D.J. and Mapp, H.P. (1995) Farm-level economic analysis incorporating stochastic environmental risk assessment. *American Journal of Agricultural Economics* 77, 8–19.

Ten Brink, B. (1991) The AMOEBA approach as a useful tool for establishing sustainable development? In: Kuik, O. and Verbruggen, H. (eds) *In Search of Indicators of Sustainable Development*. Kluwer Academic Publishers, Dordrecht, The Netherlands.

Ter Steege, M.W., Terwan, P. and Buys, J.C. (1996) *Beloning van Agrarisch Natuurbeheer in Waterland*. Centrum voor Landbouw en Milieu, Utrecht, The Netherlands.

Terwan, P. and Van der Bijl, G. (1996) Dutch country report for the inventory on landscape and nature conservation. In: Umstätter, J. and Dabbert, S. (eds) *Policies for Landscape and Nature Conservation in Europe*. Department of Farm Management, University of Hohenheim-Stuttgart, Germany.

Udo de Haes, H.A., Tamis, W.L.M., de Snoo G.R. and Canters, K.J. (1993) Algemene Natuurkwaliteit – Een prima idee maar het moet eenvoudig blijven. *Landschap* 10(4), 53–58.

Udo de Haes, H.A., de Snoo, G.R., Tamis, W.L.M. and Canters, K.J. (1997) Algemene natuurkwaliteit: soortenrijkdom in relatie tot grondgebruik. *Landschap*, 14(1), 47–51.

Van der Windt, H.J. (1996) Kwantificeren van de natuur, al een eeuw lang omstreden. In: Strijker, D. and Sijtsma, F.J. (eds). *Meten van Natuur*. Stichting Ruimtelijke Economie, Groningen, The Netherlands.

Van Harmelen, W., Kruk, M. and Mugge, F.C.T. (1995) *Een snip voor een grutto: meten, opgeven en controleren van weidevogels bij natuurproduktiebetaling.* Milieubiologie, Rijksuniversiteit Leiden, Leiden, The Netherlands.

Van Paassen, A.G., Terwan, P. and Stoop, J.M. (1991) *Resultaatbeloning in het Agrarisch Natuurbeheer.* Centrum voor Landbouw en Milieu, Utrecht, The Netherlands.

Veeneklaas, F.R., van Eck, W. and Harms, W.B. (1994) *De Twee Kanten van de Snip over Economische en Ecologische Duurzaamheid van Natuur,* Report 351. DLO-Staring Centrum, Wageningen. The Netherlands.

Verhoeven, J.T.W., Wossink, G.A.A. and Reus, J.A.W.A. (1995) An environmental yardstick in farm economic modelling of future pesticide use; the case of arable farming. *Netherlands Journal of Agricultural Science* 42, 331–341.

Wahlberg-Jacobs, B. (1991) *Integration von Naturschutz in die landwirtschaftliche Praxis – Vorgestellt anhand der Verträglichkeitsanalyse.* Verlag Dr. Kovac, Hamburg, Germany.

Wossink, G.A.A., de Koeijer, T.J. and Renkema, J.A. (1992) Environmental economic policy assessment: a farm economic approach. *Agricultural Systems* 39, 421–438.

Indicators for High Nature Value Farming Systems in Europe

9

David Baldock

INTRODUCTION

The term 'high nature value farming systems', or 'HNV systems', has come into use relatively recently. It may never have been defined precisely, and is likely to be difficult to specify in quantitative terms, but it has found a role in policy discussions, especially in relation to agri-environment measures. Before considering whether indicators can be identified and used for an emerging concept of this kind, it is pertinent to review the context in which the term is being deployed and the relevance of trying to identify farming systems of this kind.

THE REDISCOVERY OF HIGH NATURE VALUE FARMING

There is nothing new about HNV systems, many of which consist of rather traditionally managed farms. At one level, an explanation for the growing use of the term can be found in scientific work and the development of environmental thought and policy. There is increasing recognition of the important contribution made by some farming practices and systems to the maintenance of valued semi-natural habitats and landscapes in Europe. This has fuelled concern about the fate of these farming systems, many of which appear to be in decline and in danger of intensification or abandonment (Beaufoy *et al.*, 1994). Underlying this new interest is recent research and the growing appreciation of the importance of the 'wider countryside' as a habitat for many species of conservation interest (Bignal and McCracken, 1996; Pain and Pienkowski, 1997). For birds in particular, a growing body of research at a European scale has confirmed this impression.

©CAB INTERNATIONAL 1999. *Environmental Indicators and Agricultural Policy* (eds F.M. Brouwer and J.R. Crabtree)

The counterpart to this conservation debate has been a new emphasis on the social and environmental benefits of farming within the agricultural community. In a changing political climate there has been a concern to draw attention to the often positive role of agriculture in managing a sizeable proportion of the European landscape and to balance criticisms of the environmental cost of contemporary production systems.

These two strands of debate have been reinforced by advances in European Union (EU) policy which have required greater scrutiny of the sometimes ambiguous role of agriculture in environmental stewardship. Both environmental and agricultural policy initiatives have underlined the significance of the contribution of farming to nature conservation in Europe and have given new impetus to work that seeks to improve understanding of the relationship between farming and wildlife.

In the environment domain, legislative measures such as the EU Birds and Habitats Directives have required a programme of site identification, the formulation of management plans and the framing of mechanisms to protect and enhance sites designated as being of European importance. Scientific criteria are used to identify species and habitats of importance in conservation terms. It has become clear that a significant proportion of semi-natural habitats, which Member States are obliged to protect under this legislation, have been and continue to be managed by farmers, including grassland, other grazed areas and even some arable land. Some farmers have a central role as managers of nature, even though others have been responsible for large-scale destruction.

In agricultural policy, the reform of the Common Agricultural Policy (CAP) in 1992 led to the emergence of a significant new strand of agri-environment measures which became a compulsory element in European policy for the first time. This generated new sources of funding for providing farmers and other land managers with incentives for environmental management over a significant area, including sites of high nature value. The EC agri-environment Regulation 2078/92 makes available funding from the CAP budget, Fond Européen d'Orientation et de Garantie Agricole (FEOGA), for schemes developed by Member States, both for positive management of the countryside and for measures to reduce pollution. Generally, half the cost of measures implemented by Member States can be met from the CAP budget. However, the proportion rises to 75% in the least developed regions of the Community, classified under the EU regional policy as 'Objective 1'. By 1996 the FEOGA budget for the agri-environment Regulation had risen to around ECU 1.4 billion and total expenditure on agri-environment payments, including national contributions, was considerably higher than this (DG VI, 1997).

Looking further ahead, it is likely that the EU budget for agri-environment measures will be enlarged after 1999 and Member States will be under increasing pressure to refine their programmes so as to secure more precisely defined environmental benefits. This intention is spelled out briefly, but unambiguously, in the European Commission's synopsis of its proposals for the future of the CAP embodied in *Agenda 2000* (CEC, 1997a). At the same time, there is expected to be increasing international examination of existing mechanisms for agricultural support in the EU and elsewhere as the next phase of World Trade Organisation (WTO) negotiations on agricultural trade approaches. Over time, production-related supports are likely to be reduced or perhaps phased out entirely. By contrast, incentives for farmers to produce 'public goods', such as positive landscape and nature conservation benefits, should be acceptable within the framework of WTO negotiations for the foreseeable future. 'Decoupled' payments with environmental objectives are acquiring a strategic importance in agricultural policy which was not apparent prior to 1992.

If payments to farmers for providing environmental services are assuming a larger role in both agriculture and nature conservation policy, this raises fundamental questions about the type of rural environment that Europeans consider desirable. Which forms of land management generate real environmental benefits and justify payments to farmers? If we want to maintain certain habitats for conservation purposes, must we create large-scale nature reserves, or can we rely on appropriate forms of agriculture?

There is a good measure of consensus about maintaining the cultural landscapes which are so characteristic of much of the European countryside. It is more challenging to distil precise nature conservation objectives which can be used to identify land management goals. However, the development of a more systematic approach to nature conservation in recent years is providing a foundation for determining how farms need to be managed for the benefit of a range of semi-natural habitats and the wildlife associated with them. Increasingly, it is possible to develop targets for individual species, habitats and wider ecosystems and to derive management plans from these. This is a more sophisticated procedure than relying on the establishment of protected areas and conferring protection on rare species. The new approach has important implications for the management of agricultural land.

In the 15 member states of the EU, there is little wilderness remaining and many species depend on semi-natural habitats, a large proportion of which are subject to some form of agricultural or woodland management. Given the maintenance and enhancement of the nature conservation value of these habitats as a critical objective, it is then important to identify those agricultural practices and farming systems that have been associated with their emergence and management. Because of this association, such farming systems and practices can be considered of high nature value. Perhaps, more accurately, they can be said to have created a matrix of physical conditions in which valued species have a good chance of achieving a favourable status. Over time, it may be possible to generate equally favourable conditions for these species by other forms of land management, such as more extensive networks of nature reserves than now exist or seem feasible in the short term. However, for the moment there is a clear linkage between the farming systems themselves and the maintenance of habitats of conservation importance.

In short, an improved understanding of the relationship between nature conservation and farming systems is not only of theoretical interest but will also allow more informed land management choices. If we can identify 'high nature value farming systems', it may assist the development of both conservation and agriculture policies, including the targeting of agri-environment measures. There is an urgency to the task since many of the farming systems so integral to the maintenance of Europe's cultural landscapes and semi-natural habitats generate relatively low incomes and face an uncertain future. Their vulnerability to technical, social and economic change requires us to consider ways either of maintaining existing forms of management in the face of adverse trends or developing alternative means of conserving the conservation interest of large tracts of land in Europe.

HNV SYSTEMS

The term 'high nature value farming systems' is not always used in a consistent way and on occasions refers to certain land uses rather than farming systems. In a narrow sense, it refers specifically to identifiable farming systems such as transhumance or traditional forms of dryland arable farming. However, more often, it applies to those farming activities associated with valued semi-natural habitats, even if they are not strictly 'systems' in agronomic terms. In the latter sense, they can be summarized as those forms of management which maintain important landscapes and habitats, both on the cultivated or grazed area, including semi-natural grasslands and extensive dryland cereal areas, for example, and in associated features such as hedgerows, ditches, ponds, trees and patches of small woodland which historically were managed alongside, or as part of, the farming system (Baldock and Beaufoy, 1993). The European Commission leans towards a similar characterization in a recent document on rural development in the EU (CEC, 1997b). However, there is no definitive inventory of such systems in Europe. Most of the systems identified in the literature can be described as low intensity.

In practice, a helpful way of trying to identify HNV farming systems is to establish the range of differing agricultural land uses and habitat types with which they are associated. These can be grouped in some broad categories:

- Sizeable areas of semi-natural vegetation which have been subject to relatively consistent low-intensity management in the past, such as alpine grassland and dry grassland, which is often used as pasture by sheep and other livestock, including beef cattle, goats and horses.
- Mosaic agricultural habitats made up of different land uses, including parcels of farmland with different crops, patches of grassland, orchards, areas of woodland and scrub.
- Areas of grassland and neighbouring arable land of particular importance for certain species, including birds, the habitat requirements of which are relatively well studied. Some of this farmland comprises rather intensively managed areas of, for example, wet pasture in the western Netherlands

used by wader species such as the black-tailed godwit *(Limosa limosa)*. Certain bird species will tolerate, or even benefit from, habitats found on productive, relatively intensively managed farmland where there is little botanical diversity but high-yielding crops are still compatible with feeding or breeding conditions.

- Fringe and linear habitats found within the farmed environment, such as streams, banks, field margins, hedges, ponds, etc.
- Other 'in-field' habitats where the form of management is consistent with requirements of particular species. Hares, for example, may be dependent on arable fields under certain conditions.

A wide range of farming systems is associated with such habitat types. Many are livestock systems, including large tracts of unimproved pasture and meadow in the uplands and mountains. Livestock in Europe, especially sheep, hardier cattle and, in parts of the Mediterranean, goats, do not graze exclusively on grass and contemporary forage crops but exploit a range of other habitats such as moorland, heath and Mediterranean scrub. In some regions, where pastures and meadows retain conservation interest, especially in the mountains, dairy farms are the traditional form of land management. In most of Europe, arable farming has been intensified to the point where it can no longer be described as high nature value, but there are some areas where this is not the case, especially low-yielding dryland systems retaining a sizeable proportion of fallow. Certain forms of mixed farming can be classified as HNV, notably the agro-pastoral *dehesa* system in Spain and the similar *montados* in Portugal, which combine mature trees with different forms of livestock and sometimes arable production as well. Permanent crops, including orchards, olives and vines, comprise a sizeable element of production in many regions, especially in southern Europe, and include a proportion of HNV systems, typified by older, botanically rich orchards and traditionally managed olive groves.

While there is no reliable account of the exact distribution of these systems in the EU, the majority are likely to be found within the Less Favoured Areas (Beaufoy *et al.*, 1994; CEC, 1997b). Indeed, many are marginal in economic terms.

POTENTIAL INDICATORS

There are several ways of approaching the selection of indicators for identifying HNV farming systems in Europe. Three different alternatives are considered here:

1. Taking environmentally protected areas as a proxy for conservation interest and focusing on agricultural land within these designated areas.
2. Identifying the precise land-management requirements of a range of species and habitats. For example, this might start with an analysis of those semi-natural habitats considered to be of European importance and the

requirements of individual species of conservation importance and working towards a complete inventory of the land-use management implications, including the area considered necessary to maintain the target population of each species.

3. Focusing on low-intensity farming systems, which are more readily identified and are likely to constitute much the largest element in any inventory of HNV systems.

Protected Areas

The first, and potentially the simplest, approach is to take protected areas of countryside in Europe as the primary indicator of the potential location of HNV agriculture. In principle, this is a means of identifying stretches of land selected as valuable for the conservation of landscapes or nature by public authorities. Many categories of protected land include significant areas under agricultural management, with the exception of some more strictly defined nature reserves where any form of economic activity is prohibited.

The great majority of protected areas are established under national, regional or even local legislation. A network of sites to be protected under the EU Birds and Habitats Directives, to be known as Natura 2000, is in the process of being established. This will include only sites of European importance, identified by means of fairly strict criteria laid down in the two directives. The resulting network may cover 10–15% of the land area in some Member States but will be much smaller than the full range of sites established under national and sub-national law. As a result, the European network, Natura 2000, is too narrowly defined to be an appropriate screen for selecting HNV systems.

The level of protection offered to sites under national and sub-national legislation varies greatly and, in many cases, 'normal' agricultural activity is permitted. It is necessary to distinguish between formally protected areas, even if they have little legal status, and stretches of agricultural land which have been designated as 'environmentally sensitive' but which have no legal protection at all. Environmentally Sensitive Areas (ESAs) have been created in several member states, including the UK and Denmark, as a means of targeting agri-environment policy measures on particular stretches of land. Typically, these have been selected because of their landscape, nature conservation or historic interest, or because water bodies in the zone are vulnerable to pollution from agricultural sources. Very often, ESAs will consist largely or wholly of HNV systems, but this is not necessarily the case and many Member States with large areas of land of high nature value have not established ESAs at all, preferring to rely on other means of targeting agri-environment measures.

In practice, there are considerable difficulties with relying on formally protected areas as a filter for locating HNV systems. One major concern is that there is little consistency within Europe in the designation of these areas. National and, in some cases, regional authorities have developed their own systems of protected area, with or without reference to the broad categories identified by IUCN at an international level. Six categories of this kind are shown in Table 9.1. For any one category such as 'National Park' the criteria adopted for designation tend to vary significantly between countries. The predominant categories used in Europe are Category V, the protected landscape, accounting for about two-thirds of the total, and Category IV, the habitat/species management area, which accounts for a further 19% (IUCN, 1994).

It is often difficult to establish precisely how much agricultural land falls within protected areas. One country with good data is France. Of seven main categories of protected areas in France, two contain less than 5% of agricultural land, a further two contain less than 15%, two contain around 20% and one, the *parcs naturels régionaux*, contains around 45% (IFEN, 1997). This is an interesting case because there is a total of around 5.1 million ha of agricultural land in the *parcs naturels régionaux* as opposed to approximately 1.5 million in the other six categories combined. It is not easy to identify an equivalent category of protected area in many other European countries.

The conservation requirements for this particular category of protected area are not very tightly defined. However, a comparison can be made with the area of farmland found within the more scientifically delineated *zones naturels d'intérêt écologique, faunistique, et floristique* (ZNIEFF). This is an inventory of sites of national importance for a wide range of flora and fauna following scientific criteria, and established by relatively detailed survey work at a regional and local level. This indicated that around 1.06 million ha of agricultural land lie within core sites of particular importance (ZNIEFF 1) and around 4.14 million ha in the second, but still important tier, the ZNIEFF 2 (IFEN, 1997).

Often the selection criteria for protected areas are not very precise or are not applied systematically or consistently within a country. Furthermore, the process of selecting areas can, and often does, stretch over a very long period, as has occurred with the selection of national parks in Italy. As a result, many

Table 9.1. The six IUCN management categories for protected areas (source: IUCN, 1994).

Category I	Strict nature reserve/wilderness area – protected area managed mainly for science or wilderness protection
Category II	National park – protected area managed mainly for ecosystem protection and recreation
Category III	Natural monument – protected area managed mainly for conservation of specific natural features
Category IV	Habitat/species management area – protected area managed mainly for conservation through management intervention
Category V	Protected landscape/seascape – protected area managed mainly for landscape/ seascape conservation and recreation
Category VI	Managed resource protected area – protected area managed mainly for the sustainable use of natural ecosystems

sites meeting the formal criteria laid down in legislation have not been adopted as protected areas and it is unclear whether they will be in future. In short, the series of protected areas is still incomplete, even if we accept the current set of legislative measures that underpin them as authoritative. In a recent review of protected areas in Europe, IUCN has commented that, despite recent expansion in the extent of protected areas, many European countries still have 'large additional areas of natural or semi-natural vegetation, rich in biodiversity, much of which should be within protected areas' (IUCN, 1994).

In many countries it can be observed that the tradition of site protection is to concentrate on relatively small areas, often excluding wider tracts of farmed countryside. Larger protected sites are often in the most mountainous regions where forest and unmanaged habitats frequently dominate the landscape. Indeed, these are countries where protected areas are selected in such a way as to exclude many important HNV farming systems because they are not sufficiently natural.

For all these reasons, protected areas are not satisfactory as the sole indicator of the location of HNV farming systems. None the less, they can be used together with other indicators, bearing in mind their limitations discussed above and inconsistencies between countries.

Indicators Based on Species and Habitat Management Requirements

In principle, it is possible to adopt a more scientific approach, rather than relying on the current network of protected areas. It is known that many species rely on habitats under some form of agricultural management. Recent work on bird species, which have been studied in greater detail than other fauna and flora, suggests that, at a European scale, agricultural habitats have the highest overall species richness of any category of habitat (Tucker, 1997). Over time, it should be possible to build up a detailed picture of the specific requirements of individual species which utilize farmland for part, or all, of their life cycles. This inventory could provide information on the appropriate form of land management and the agricultural systems likely to bring about such management. It might also include some estimate of the area of land required under a particular form of management in order to create a reasonable probability that a species can be maintained at a target population level in Europe. A further step would be to provide some indication of the most appropriate location for farmland managed in accordance with conservation criteria.

This kind of approach would permit a clearer understanding of which farming practices benefited which species and would provide a yardstick against which to judge how many species could potentially benefit from particular farming practices. Our understanding of these factors at a regional and European scale will increase over time with assistance from the growing volume of research work now under way. However, at present, there are insufficient data available from which to construct a detailed inventory of species requirements on farmland or to identify the precise value of different contemporary farming systems with absolute confidence.

Work by BirdLife International provides a helpful starting point for a scientific analysis of this kind. Their division of farm habitats into different categories is itself revealing. In drawing up a conservation strategy for agricultural habitats, they distinguish between eight subdivisions (Tucker, 1997):

- arable and agriculturally improved grasslands;
- steppic habitats, which are mainly dry grassland and extensive cereal crops;
- wet grassland;
- moorland;
- alpine grasslands, above the tree line;
- orchards, olive groves and other perennial crops;
- pastoral woodlands, including *dehesas* and *montados*;
- rice cultivation.

Work by BirdLife is focused particularly on a group of bird species of conservation concern in Europe. It points to the importance of low-intensity agricultural habitat types, including steppic habitats and alpine grassland. However, it also suggests that:

> a wide category of farming types, including some moderately intensive cultivations on agriculturally improved grasslands, should also be regarded as of high conservation importance. Over all, this wide range of agricultural types has the highest species richness and, more importantly, the highest number of priority species and species with an Unfavourable Conservation Status, a significant number of which have the majority of their populations within the habitat. Only the most intensively farmed habitats have an extremely impoverished bird fauna and a generally low natural value.

(BirdLife International, 1996)

Rather than considering the full suite of species and habitats of conservation interest or even those protected under European legislation, it may be possible to rely instead on a much smaller selection of indicator species. This possibility is discussed further in Chapter 7 of this volume.

Low-intensity Farming Systems

A third approach is to use agricultural and land-use data in an attempt to identify low-intensity agricultural systems. These are systems which are low in their use of external inputs, especially fertilizers and agrochemicals. They include livestock systems, cropping systems, permanent crop systems and mixed systems.

In principle, it is possible to identify many of the typical characteristics of low-intensity farming systems; in Table 9.2 some of the features normally associated with livestock and crop systems are identified separately. In both cases, low intensity implies a significantly below-average use of inorganic fertilizers and agrochemicals. Another, almost universal, characteristic is low output per hectare as measured by crop yields or weight of livestock products. The value of output is often relatively low as well, although there are examples of products able to attract a substantial premium, at least partly offsetting the limited physical output.

Many low-intensity systems still use more traditional forms of production. For example, mechanization tends to be less advanced on low-input livestock farms than on more high-yielding holdings. Many hardier regional breeds of livestock, usually well adapted to maintenance on coarser semi-natural vegetation, are concentrated on low-input farms. Often it is the severity of physical constraints on production and the absence of severe environmental intervention, such as drainage and irrigation, which oblige farmers to limit their use of inputs and in these conditions traditional practices are more likely to be retained. Where soils and climatic conditions have allowed intensification to proceed, it has done so in much of Europe.

There are several reasons for thinking that this particular category of agricultural systems will contain significant areas of high nature value. The type of characteristics which, in principle, benefit biodiversity include:

Table 9.2.   Typical characteristics of low-intensity systems.

Low nutrient inputs Low output per hectare	
Livestock systems	Crop systems
Low nutrient input, predominantly organic	Low nutrient input, predominantly organic
Low stocking density	Low yield per hectare
Low agrochemical input	Low agrochemical input (usually no growth regulators)
Little investment in land drainage	Absence of irrigation
Relatively high percentage of semi-natural vegetation	Little investment in land drainage
Relatively high species composition of sward	Crops and varieties suited to specific regional conditions
Low degree of mechanization	Use of fallow in the crop rotation
Often hardier, regional breeds of stock	Diverse rotations
Survival of long-established management practices, e.g. transhumance, hay-making	More traditional crop varieties
Reliance on natural suckling	Low degree of mechanization
Limited use of concentrate feeds	Tree crops, tall rather than dwarf – not irrigated
	More 'traditional' harvesting methods

- a low level of nutrient inputs, both historically and under current management;
- a higher proportion of semi-natural vegetation than on most farmland, including permanent pasture, *maquis*, hedgerows, etc.;
- absence or low use of agrochemicals, benefiting a wide range of fauna and flora;
- a tendency to retain a number of 'traditional' management practices, such as the late harvesting of meadows and arable crops and the use of shepherds rather than 'ranching' techniques;
- greater continuity in management practices, providing many species with periods of relative stability, facilitating adjustment to prevailing conditions (Beaufoy *et al.*, 1994).

It is clear that significant areas of HNV agricultural land are not low intensity, as discussed above. Conversely, there are expanses of low-intensity farming which are not of high nature value and even low stocking densities, less than 0.5 livestock units per hectare, can be excessive in some locations. None the less, the essential character of low-intensity farming systems implies less drastic interventions in the natural environment than in conventional high-yielding systems and provides a clear rationale for believing that a sizeable proportion of HNV systems will fall into this category.

Several of the attributes of low-intensity farming can be defined more easily than those for HNV systems and there is a variety of agricultural data available at a national and European scale of potential value in the search for indicators. Several sources of information for identifying low-intensity systems are shown in Table 9.3 (Beaufoy *et al.*, 1994). Some potential indicators, such

Table 9.3. Potential sources of information for identifying low-intensity agricultural systems.

Land use and vegetation types	Indicators of agricultural management intensity	Agricultural input/output indicators	Wildlife indicators (ecological requirements)
Proportion of certain vegetation types, e.g. *maquis*, moorland, steppe, pseudo-steppe, permanent grassland	Irrigation Drainage Fallow periods Breed of livestock Crop varieties and associations Age, size of trees (in orchards and olive groves)	Level of use of: fertilizer and manure (phosphate and nitrogen) pesticides herbicides Stock density Value of output per hectare Employment (full time, part time)	Species and habitats of European, national and regional importance 'Indicator species', for example, from the Birds and Habitats Directives

as stocking densities, level of fertilizer and manure use and agrochemical use, are susceptible to measurement and are the subject of regular surveys in most Member States. None the less, many of these indicators must be treated with great caution. Some potential indicators are only relevant to a very narrow range of systems. Agricultural statistics rarely provide much information on the nature of the farming systems under which land is managed. Data on input use are often highly aggregated, making it difficult to identify HNV systems at a local level. Stocking density statistics are useful but are not always a reliable guide to the intensity of a pastoral farming system. It is essential to have information about the natural conditions in the area concerned as well as the number of livestock present and the pattern of grazing.

Three of the potentially most useful indicators of low-intensity farming systems are described below.

The proportion of semi-natural vegetation found on the farm
This will vary depending on the region concerned. It may consist of permanent pasture, wood pasture, *maquis* or similar types of dry shrubby habitats. Semi-natural vegetation tends to be displaced as intensification proceeds and semi-natural pasture is transformed into drained and re-seeded, high-yielding grassland, for example. Typically, the persistence of semi-natural vegetation on a significant scale indicates the absence of agricultural land improvement, arterial drainage and similar works. As well as having habitat value in its own right, it suggests that fertilizers have not been applied to any significant degree. While this is a useful indicator, obtaining reliable data on a national or European scale is difficult.

Stocking densities
A stocking density index provides a measure of the average number of stock in a region or on a farm during the course of the year. However, the number may vary considerably between seasons or years and stocking density indices give only a limited guide to the management practices in the region. The methods of managing both stock and forage have a major bearing on the resulting pattern of vegetation and the value of the land as habitat for different species.

Stocking density statistics are one of the most direct and helpful indicators of intensity in livestock systems. However, their interpretation requires considerable care. Figures that include non-ruminants give an impression of overall pressure on the environment from livestock wastes and indicate intensity as a whole, but they may not be useful in pinpointing the density of grazing ruminants, which is often a major concern. Even where non-ruminants are excluded, average stocking density figures combine livestock which are permanently or largely housed in buildings with those that are predominantly outdoors and grazing. Even in this case they are not always a reliable guide to the intensity of a pastoral system. Whether a given stocking density can be considered 'low intensity' depends on the natural conditions and forage available in the area concerned. Ecologically 'appropriate' stocking densities may be difficult to define other than at a very local scale and a programme of monitoring and research may be needed to refine first impressions.

The method of computing grazing livestock units (LUs) varies significantly between countries. Within the EU, many member states have their own conventions for domestic purposes, which do not necessarily coincide with the system used for EU legislation. The EU system denotes a ewe as 0.15 LU, regardless of breed, while in France there is a range of 0.13 to 0.18 LU (Julien, 1991), and in the UK, 0.06 to 0.11 (MAFF, 1977). In practice, forage consumption varies significantly between species and breeds, and a differentiated LU conversion ratio may be more appropriate. In Britain, hardy hill breeds of ewe, such as Herdwick or Ronaldsay, are considered as equivalent to 0.1 LU, whereas larger, more productive lowland breeds are counted as 0.15 LU (Nix, 1993).

Similar variations in the size, productivity and forage consumption of other livestock types can be noted. There may also be significant differences in the way in which forage hectares are calculated. For example, in France some statistics on the total forage area of regions do not include communal or public grazing land. Given the existing variations in methods of measurement, as well as in the dates of surveys and the scale of geographical units for which data have been collected, comparisons between countries must be undertaken with great caution.

Levels of inorganic fertilizer and agrochemical use
These can be considered as direct indicators of low-input farming and are readily quantifiable. Ideally such data are required at the field or farm level but are often available only at a regional or more aggregated level. Such figures will mask important differences between farming systems, individual farms and fields. Data on nutrient inputs are particularly useful but tend to show applications from inorganic fertilizers, rather than from all sources, including livestock wastes and aerial deposition.

For these and other potentially useful indicators, such as agricultural land use and farm management data, there are limitations on the information available at a European level. Often information is obtainable at a level that is too highly aggregated to be useful and there are problems of consistency, incompleteness, lack of recent survey work, difficulties of interpretation, etc. Furthermore, there does not appear to be any single indicator which is adequate for identifying the extent and location of low-intensity agricultural land. Approaches based on agricultural statistics and a limited range of criteria, such as land cover, input use, average stocking densities, etc., inevitably provide an incomplete picture. Local patches of low-intensity land are likely to be overlooked and the details of current farming practices will not be revealed by such an approach. Ground surveys, aerial photography and detailed local land use maps are potentially very valuable for locating smaller areas.

Expert knowledge is an essential complement to any data-based system, particularly as it is often the only way to gain an insight into the functioning of farming systems. By combining information from a variety of sources with expert knowledge, it is possible to establish a reasonable picture of the distribution of low-intensity systems in most European countries. Inevitably, the results are only approximate and there are unavoidable inconsistencies between countries. None the less, this pragmatic approach provides a useful foundation for policy discussion and is a workable means of identifying a significant proportion of low-intensity farmland in Europe.

CONCLUSION

Each of the three approaches outlined here confirms the availability of information sources and data sets which provide insights into the character of HNV farming systems and contribute towards the process of identifying their location. Some of the information available at a local, or even national, level is particularly useful and, in combination with expert knowledge, permits initial judgements to be made about the types of farming most associated with high nature value and their geographical distribution. At a European level, the diversity and variability of the information and inconsistencies between regions have become a more serious impediment, but the information still can be used in a pragmatic way, particularly in combination with expert knowledge. In the absence of appropriate and reliable indicators, it is not possible to create a satisfactory inventory of HNV farming systems in Europe. More limited goals can be achieved, however. For example, the concept of HNV farming can be developed and debated, preliminary mapping exercises can be undertaken and some input can be made to policy debates. If, as seems likely, considerable resources are required to generate the information necessary to define HNV systems more precisely, the pragmatic use of available information has value as an interim measure.

REFERENCES

Baldock, D. and Beaufoy, G. (1993) *Nature Conservation and New Directions in the EC Common Agricultural Policy.* Institute for European Environmental Policy, London.

Beaufoy, G., Baldock, D. and Clark, J. (1994) *The Nature of Farming: Low Intensity Farming Systems in Nine European Countries.* Institute for European Environmental Policy, London.

Bignal, E.M. and McCracken, D.I. (1996) Low-intensity farming systems in the conservation of the countryside. *Journal of Applied Ecology* 33, 413–424.

BirdLife International (1996) *Nature Conservation Benefits of Plans under the Agri-Environment Regulation (EC) 2078/92.* BirdLife International, Sandy, Bedfordshire.

CEC (1997a) *Agenda 2000: For a Stronger and Wider Union*. Commission of the European Communities, Brussels.

CEC (1997b) *Rural Developments, CAP 2000 Working Document*. European Commission DG VI, Brussels.

DG VI (1997) Data on Regulation 2078/92 published in House of Commons' Agriculture Committee 1997, Second Report, Session 1996–97, *Environmentally Sensitive Areas and Other Schemes under the Agri-environmental Regulation*. HMSO, London.

IFEN (1997) *Agriculture et Environnement: les Indicateurs*. Institut Français de l'Environnement, Orléans, France.

IUCN (1994) *Parks for Life: Action for Protected Areas in Europe*. IUCN/FNPPE/WWF/WCNC/BirdLife, Gland, Switzerland.

Julien, M. (1991) *L'extensification des Productions d'Herbivores à la Lumière du RGA 1988*. Ministère de l'Agriculture et de la Fôret, DERF/SCEES/UNIGRAN, Comité National d'Extensification et de Diversification, Paris.

MAFF (1977) *Definition of Terms used in Agricultural Business Management*. HMSO, London.

Nix, J. (1993) *Farm Management Pocketbook*, 23rd edn. Wye College, University of London.

Pain, D.J. and Pienkowski, M.J. (eds) (1997) *Farming and Birds in Europe: The Common Agricultural Policy and its Implications for Bird Conservation*. Academic Press, London.

Tucker, G. (1997) Priorities for bird conservation in Europe: the importance of the farmed landscape. In: Pain, D.J. and Pienkowski, M.J. (eds) *Farming and Birds in Europe: The Common Agricultural Policy and its Implications for Bird Conservation*. Academic Press, London, pp. 79–116.

Agri-environmental Indicators for Extensive Land-use Systems in the Iberian Peninsula

10

Begoña Peco, Juan E. Malo, Juan J. Oñate, Francisco Suárez and José Sumpsi

INTRODUCTION

Considerable attention has been paid in the past decade to the use of agri-environmental indicators for evaluating the state and environmental value of farming systems and the consequences of agricultural policies. There is, however, still no consensus about which indicators should be used.

Some authors have proposed indicators that focus on socio-economic aspects linked to farming activity and the problems of marginalization (e.g. Baldock *et al.*, 1996). Others have placed more emphasis on environmental aspects, stressing the problems caused by intensive farming techniques (e.g. McKenzie *et al.*, 1992; Ministerio de Medio Ambiente, 1996; OECD, 1997). To our knowledge, however, the methodologies so far developed have not included, or have not specified in sufficient detail, how they might be applied to extensive systems or to the issue of land abandonment.

In contrast to the central and northern regions of the European Union (EU), a very large area of the Iberian Peninsula is used for extensive farming (Beaufoy *et al.*, 1994). In general terms, the agrosystems are characterized by dry management, without any input of water other than scant and irregular Mediterranean rainfall (Suárez *et al.*, 1997b). Their other key characteristic is their diversity, including both well-defined crop uses (e.g. dry cereal farming in the centre of the peninsula or olive groves), and extensive livestock husbandry (e.g. *dehesas*, grazing and hay meadows, etc.). Many of these agrosystems have extremely high natural and conservation value, with numerous internationally recognized endangered habitats and/or species, listed in several 'Red Books' and Annexes of EU conservation directives (Viedma and Goméz-

Bustillo, 1985; Blanco and González, 1992; De Juana *et al.*, 1993; MOPTMA, 1995; Domínguez Lozano *et al.*, 1996).

Dry cereal farming and tree-covered *dehesas* are notable amongst these agrosystems for their natural value. The former are outstanding due to their bird communities, with endangered species of international importance, e.g. the great bustard, *Otis tarda* (Alonso and Alonso, 1990a) or the little bustard, *Tetrax tetrax* (Martínez, 1994), and species which in the EU context are distributed, or have their main breeding and wintering populations, in the Iberian Peninsula, e.g. the sand grouse *(Pterocles orientalis)* and the pin-tailed sand grouse *(P. alchata)* (Suárez *et al.*, 1997a) or Dupont's lark, *Chersophilus duponti* (Garza and Suárez, 1990).[1] The value of the vegetation is also high, with many endemic taxa and notable communities from a botanical and biogeographical perspective, such as grasslands interspersed amongst crops or stands of rushes and meadows in groundwater discharge zones (Bernáldez *et al.*, 1989; Rey Benayas *et al.*, 1990). The limnological value of the temporal wetlands associated with discharge zones has also been highlighted (Florín *et al.*, 1996).

The tree-covered *dehesas* have been considered by numerous authors to be amongst the most environmentally valuable agrosystems on the Iberian Peninsula (see Campos, 1993). They host extraordinarily diverse communities of vertebrates (Díaz *et al.*, 1997), with numerous species endangered at a European level, e.g. the black vulture, *Aegypius monachus* (Donázar, 1993) and black stork, *Ciconia nigra* (Tucker and Heath, 1994), and one species endangered in world-wide terms, the imperial eagle, *Aquila adalberti* (González *et al.*, 1990). They are also extremely important as wintering habitats for the crane, *Grus grus* (Alonso and Alonso, 1990b). The value of *dehesas* is not restricted to vertebrates but also includes an extremely high diversity of vascular plants (Marañón, 1985; Peco, 1989) and butterflies (Viejo *et al.*, 1989).

In these two systems, the Common Agricultural Policy (CAP) has caused a dual process of intensification and abandonment which has not been uniform at a regional scale or, in the case of the *dehesas*, at a local level either. This spatial duality of the process can have a drastic impact on the continuity of agrosystems, hence fragmenting the optimum habitat of numerous species (González Bernáldez, 1991).

For all of the above reasons, the dry cereal steppes and *dehesas* provide excellent case studies for testing the comparative definition and application of indicators whose purpose is to measure the environmental health of extensive agrosystems. This chapter presents a list of agri-environmental indicators and evaluates their suitability for measuring the effects of the CAP on the environment, landscape and nature in these two paradigmatic agrosystems. The discussion of the results allows an evaluation of the methodology used and an analysis of the difficulties that arise from the selection and application of the proposed system of indicators. Finally, a series of suggestions for improving the suitability of European statistical data when applied in this context is proposed.

SELECTION CRITERIA

An initial list of agri-environmental indicators was prepared, for the specific context of extensive cereal and *dehesa* systems. The indicators were designed within the framework proposed by OECD (1997) on the basis of scientific information and expert knowledge about agricultural factors that affect the conservation of the most valued species and habitats. The aim was to maximize the level of information provided by each indicator by selecting variables important in determining environmental conditions with a level of conceptual clarity and conciseness that provided unequivocal identification. At the same time, the aim was to ensure that they were easy to measure in quantitative terms (Table 10.1). This initial list included virtually all of the potentially relevant items in the functioning of the associated agro-ecosystems. Although particular emphasis was placed on steppe birds in the case of the dry cereal areas, and vertebrates and vegetation in the case of the *dehesas*, other assessable items in these ecosystems (e.g. naturalness, absence of pollution and traditional landscape) were also considered.

This initial list was analysed exhaustively and then pared down because: (i) the significance of the information provided by each indicator in the test agrosystems was variable; (ii) redundancies arose when two or more indicators reflected the same process; and (iii) suitable statistical data were not fully available for all of the indicators. These three criteria – relevance, suitability of data and non-redundancy – were applied sequentially to each indicator in the initial list, as shown in the flow chart presented as Fig. 10.1. The result was a synthesized list of indicators for each of the two agrosystems in which those with data available, whether highly (type A) or moderately (type B) relevant, were maintained, along with those of high relevance but lacking the appropriate statistical information (type C) (Tables 10.2 and 10.3). In this second list the qualitative relationship between each indicator and the environmental value of the system to which it was applied was formalized. This relationship may be one of three types:

1. positive, where a change in the value of the indicator reflects a change in the same direction of the environmental value of the system;
2. negative, where a change in the value of the indicator reflects a change in the opposite direction to the environmental value; or
3. hump-shaped, where intermediate values of the indicator show an optimum environmental value while the two extremes show minimum environmental values (Fig. 10.2).

The same classification was applied to the relationship between each indicator and four overall environmental objectives: conservation of biodiversity (species and habitats), fertile soil conservation and erosion prevention, fire prevention and pollution prevention.

Table 10.1. Initial list of indicators of the effects of CAP on environment, landscape and nature in dry-cereal farming and *dehesa* systems on the Iberian Peninsula.

Indicator	Relevance		Measurement	Problems
	Cereal	*Dehesa*		
Area of natural forest (ANF)	LR	HR	Area afforested GA^{-1}	Only regional statistics exist
Area of tree plantations (ATP)	HR	HR	Area of commercial tree plantations GA^{-1}	Only regional statistics exist
Area of crops under open woodlands (ACW)	–	HR	Area of crops amongst trees GA^{-1} In some cases, other areas for pastures GA^{-1}	–
Tree density in *dehesa* systems (TD)	–	HR	No. of trees ha^{-1}	No statistics
Area of scrub (AS)	HR	HR	Area of scrub GA^{-1}	–
Area of low scrub and barren land (ALS)	HR	MR	Area of low scrub and barren land GA^{-1} In some cases, low scrub and barren land + *esparto* grass GA^{-1}	–
Area of dry pastures (ADP)	HR	HR	Area of dry pastures GA^{-1}	Same data in statistics as AIP
Area of improved pastures (AIP)	LR	MR	Area of improved pastures GA^{-1}	Same data in statistics as ADP
Area of pasture under open woodland (APW)	–	HR	Pastures under open woodland GA^{-1}	–
Crop boundaries (CB)	HR	LR	Area of boundaries GA^{-1} Approximation: boundary length Crop area $((\text{average field size})^{0.5})^{-1}$	No statistics. May be estimated via field size
Hedgerows (H)	LR	LR	Area of hedgerows GA^{-1}	No statistics. In *dehesas* easily confused with CB if estimated via field size
Area of fallow land (AFL)	HR	LR	Area of fallow land GA^{-1} Fallow area (arable area)$^{-1}$ (kind of fallow index)	Unable to differentiate between stubble types: short/medium/ long term, ploughed, chemical

Table 10.1. *contd.*

Indicator	Relevance		Measurement	Problems
	Cereal	*Dehesa*		
Area of crops (AC)	MR	LR	(Arable area – fallow area) GA^{-1}	–
Degree of crop intensification (DCI)	HR	MR	Cereal area + dry legume area (arable area – fallow area)$^{-1}$	May be redundant with DFS and partially with AIC and AIL
Area of legume crops (ALC)	HR	LR	Legumes area GA^{-1}	Confused in *dehesas*
Diversity of agricultural substrata (DFS)	HR	LR	No. of agricultural substrata with >5% of GA (only usage of non-intensive crops)	May be redundant with DCI, AIC and AIL
Area of intensive crops (AIC)	HR	HR	Arable area A + B +. . . (GA after cataloguing crop types)$^{-1}$	Redundant with DCI and AIL
Area of irrigated land (AIL)	HR	HR	Irrigated area GA^{-1} Area of irrigated crops GA^{-1}	Redundant with DCI and AIC
Area of riparian forest (ARF)	MR	MR	Area of riverside vegetation GA^{-1}	No statistics
Area of lakes and lagoons (ALL)	MR	MR	Area of lakes and lagoons GA^{-1}	No statistics
Area of wet meadows (discharge wetlands) (AWM)	MR	MR	Area of wet meadows GA^{-1}	Normally included in statistics as AMP
Stocking density (SD)	MR	HR	LU GA^{-1} or LU UAA^{-1} Present stocking density (stocking density in 1962)$^{-1}$	Less precise than EL and IL. Second measure has no statistics
Extensive-husbandry livestock (EL)	HR	HR	LU GA^{-1} or LU UAA^{-1} Present stocking density (stocking density in 1962)$^{-1}$	Statistics do not differentiate intensive vs extensive husbandry
Intensive-husbandry livestock (IL)	HR	HR	LU UAA^{-1} Expenditure on stockfeed UAA^{-1}	Statistics do not differentiate intensive vs extensive husbandry
Type of livestock (TL)	LR	HR	LU of each species GA^{-1} LU of each species (total LU)$^{-1}$	Redundant with DEL

Table 10.1.   *contd.*

Indicator	Relevance		Measurement	Problems
	Cereal	Dehesa		
Diversity of extensive livestock (DEL)	LR	HR	No. of species with >5% of herd in LU	No statistics
Mechanization (M)	LR	LR	HP UAA^{-1} HP (arable area)$^{-1}$	May be redundant with PI
Herbicides (H)	HR	HR	kg ha^{-1}	Only regional estimations exist
N and P fertilizers (F)	HR	HR	kg ha^{-1}	Only regional estimations exist
Insecticides (INS)	HR	HR	Total kg (arable area)$^{-1}$ Total kg GA^{-1}	Redundant with H, F, AIC, AIL and PI. No statistics
Farm waste and manure (FWM)	HR	MR	kg ha^{-1}	Redundant with IL Only regional estimations exist.
Production intensification (PI)	HR	HR	Intermediate expenditure UAA GA^{-1} Intermediate expenditure on crops (arable area)$^{-1}$ Intermediate expenditure on livestock LU^{-1}	Dependent on FADN coberture, although data can only be extrapolated at regional or national level
Farming calendar (FC)	MR	MR	Dates Duration in practice (e.g. no. of days grazed)	No statistics
Fires (F)	MR	MR	Percentage of burned area	No statistics
Spatial patchiness (SP)	MR	MR	Average interconnected area of each use	No statistics
Interspersion (INT)	MR	MR	No. of changes of use per linear km	No statistics

HR, High relevance; MR, medium relevance; LR, low relevance; GA, geographic area; UAA, utilized agricultural area; LU, livestock unit; HP, horse power; FADN, Farm Accountancy Data Network.

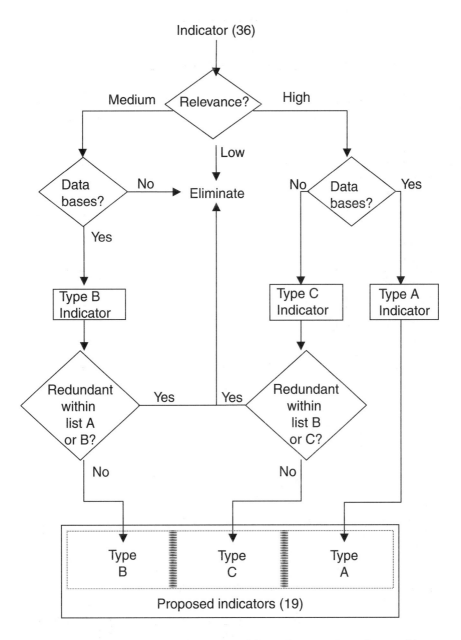

Fig. 10.1. Flow chart showing the process followed to pare down the initial list of indicators (see text for a full explanation).

Table 10.2. Relationships between selected agri-environmental indicators for dry cereal farming and the overall environmental value of the system and selected environmental issues, with reference to reflected main acting process and indicator type.

Indicator	Relationship to overall environmental value	Relationship to environmental issues				Process	Type
		Diversity conservation	Soil conservation	Fire prevention	Pollution prevention		
Area of tree plantations	Negative (aggravated if dispersed)	N	p	N	–		C
Area of scrub	Negative	N	p	N	–	Ab	A
Area of low scrub and barren land	Hump-shaped	U	p	–	–	In, Ab	A
Area of dry pastures	Hump-shaped	P	p	–	–	In, Ab	A
Crop boundaries	Positive	P	p	–	–	In	C
Area of crops	Hump-shaped	U	N	–	–	In, Ab	B
Degree of crop extensification	Positive	P	–	–	P	In	A
Area of fallow land	Positive (enhanced by management)	P	n	–	–	In	A
Area of legume crops	Positive	P	n	–	n	In	A
Area of irrigated land	Negative	N	n	–	N	In	A
Diversity of agricultural substrata	Positive (negative only possible at a small scale)	P	–	–	–	In	A
Extensive-husbandry livestock	Hump-shaped	U	–	P	–	In, Ab	C
Intensive-husbandry livestock	Negative (aggravated by poor waste management)	–	–	–	N	In	C
Production intensification	Negative	N	–	–	N	In	C

Environmental issues: N, negative; P, positive; U, hump-shaped; n, negative but less important; p, positive but less important; u, hump-shaped but less important.

Process: In, intensification; Ab, abandonment.

Indicator type: A, highly relevant and with existing suitable databases; B, moderately relevant and with existing suitable databases; C, highly relevant and without existing suitable databases.

Table 10.3. Relationships between selected agri-environmental indicators for *dehesas* and the overall environmental value of the system and selected environmental issues, with reference to reflected main acting process and indicator type.

Indicator	Relationship to overall environmental value	Relationship to environmental issues				Process	Type
		Diversity conservation	Soil conservation	Fire prevention	Pollution prevention		
Area of natural forest	Hump-shaped	U	p	N	–	In, Ab	C
Area of tree plantations	Negative	N	n	N	–	Ab	C
Area of scrub	Negative	N	p	N	–	Ab	A (C)
Area of dry pastures	Negative	U	U	–	–	In	A (C)
Area of crops under open woodlands	Hump-shaped	U	n	–	–	In, Ab	A (C)
Area of pastures under open woodland	Positive	P	P	–	–	In, Ab	A (C)
Tree density	Positive	P	P	–	–	In	C
Area of irrigated land	Negative	N	n	–	N	In	A (C)
Extensive-husbandry livestock	Hump-shaped	U	n	P	–	In, Ab	C
Intensive-husbandry livestock	Negative (aggravated by poor waste management)	–	n	–	N	In	C
Diversity of extensive livestock	Positive	P	–	–	–	In	C
Production intensification	Negative	N	–	–	N	In	C

Environmental issues: N, negative; P, positive; U, hump-shaped; n, negative but less important; p, positive but less important; u, hump-shaped but less important.

Process: In, intensification; Ab, abandonment.

Indicator type: A, highly relevant and with existing suitable databases; B, moderately relevant and with existing suitable databases; C, highly relevant and without existing suitable databases.

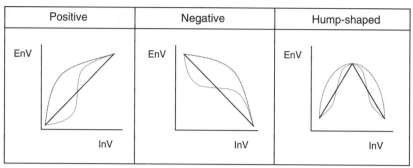

Fig. 10.2.   Possible relations between environmental value (EnV) and indicator value (InV).

THE PROPOSED INDICATORS

The initial list of possible indicators was very long at 36 (Table 10.1). Three of the indicators were exclusive to *dehesas* and the other 33 (92%) were common to both agrosystems. The degree of concordance according to the different types of relevance amongst the two agrosystems was less uniform. Eleven indicators (30%) were highly relevant in both, while seven (19%) were moderately relevant. Seven indicators (19%) were highly relevant to one of the two agrosystems but scarcely relevant to the other, two (5%) moderately and slightly relevant, respectively, and three (8%) were applicable and highly relevant only to the *dehesas*.

The availability of statistical data for the construction of the indicators is far from ideal in either of the agrosystems. In absolute terms, statistical data are available for only 46% of the dry cereal farming indicators and 47% of the *dehesas* indicators and availability at a local level is even poorer – around one-third in both systems.

There is a considerable degree of potential overlap between the 36 indicators, and thus only 21 (42%) are utterly exclusive. The rest result in some type of mutual redundancy.

The application of the methodology set out above has enabled the following synthesized list of 19 indicators to be developed:

1.   *Area of natural forest*: relatively frequent in the *dehesa*, they improve the habitat diversity for fauna. Some are included in the EEC Habitats Directive (92/43/EEC, here HD) coded as 32.11 and 45.22.
2.   *Area of tree plantations*: low ecological value and highly destructive effects for steppe birds in dry cereal farming systems. In the *dehesa*, they are mainly highly homogeneous eucalyptus plantations, of no value for flora or fauna.
3.   *Area of scrub*: in dry cereal systems these formations are the best refuge for local flora species and some are included in HD (15.18 – priority habitat, 15.19 – priority, 31.7, 32.131–136, 32.26), although the majority of those in the

dehesa have low values (e.g. *Cistus* spp.). For the fauna, they involve the entry of bird species from developed scrub with low conservation value (i.e. warblers) and the disappearance of steppe birds. In *dehesas*, they are a good indicator of abandonment (colonization to scrub). The negative effect of this variable on the rabbit results in its partial responsibility for the decline of some large predators (e.g. imperial eagle, Iberian lynx).

4. *Area of low scrub and barren land*: in dry cereal systems such areas provide shelter to successional species in transition between HD grasslands (34.5 – priority) and scrubland, and constitute a fundamental habitat for certain birds (e.g. Dupont's lark and sand grouse). Some are included in HD (15.12, 15.19).

5. *Area of dry pastures*: essential for both steppe and *dehesa* birds and included as priority habitats in HD (34.5).

6. *Area of pastures under open woodland*: decisive for *dehesa* birds and included as priority habitats in HD (32.11).

7. *Area of crops under open woodland*: important in the *dehesa* as complementary feeding and breeding habitat for birds, although in high proportions reflects intensification.

8. *Area of crops*: a measure of exploitation pressure on the environment. Crops are a habitat for weed species if herbicides are not applied heavily. A habitat for certain steppe species when cereal crops are planted (great bustard, little bustard and pin-tailed sand grouse).

9. *Crop boundaries*: in dry cereal systems, this type constitutes a refuge for successional species and weeds. Essential for steppe bird feeding and as a refuge for other vertebrates. The greater the proportion of crop area, the greater their importance.

10. *Degree of crop extensification*: a measure of extensification that can be linked to abundance of certain steppe birds in dry cereal systems.

11. *Area of fallow land*: in dry cereal systems, this type constitutes habitat for colonizing plants. It is associated with low-intensity cropping, which permits the maintenance of higher diversity. Essential for steppe birds when not cultivated.

12. *Area of legume crops*: can be linked to crop diversity and indirectly to a possible lower usage of inorganic fertilizers. It is a habitat for certain bird species (great bustard, little bustard and pin-tailed sand grouse).

13. *Diversity of agricultural substrata*: closely related to presence and density of bird species with high conservation values in dry cereal systems.

14. *Area of irrigated land*: good measure of pressure, associated with development of intensive crops. Its expansion implies the destruction of habitat for birds.

15. *Extensive livestock husbandry*: this is the valuable type of animal husbandry for vegetation maintenance, although part of the herd may be managed intensively. Basically sheep and goats and, to a lesser extent, cattle in dry cereal systems and cattle, sheep, goats and pigs in the *dehesa*.

16. *Intensive livestock husbandry*: permits evaluation of livestock density in relation to extensive livestock husbandry.

17. *Diversity of extensive livestock*: important in the *dehesa* in relation to intensive/extensive livestock dichotomy and to different uses of the landscape by different livestock species. Changes in relative importance of different species may indicate intensification and change in husbandry system. It can be assumed that a variety of herbivores can favour diversity in vegetation.

18. *Production intensification*: overall measure of general degree of intensification via intermediate expenditure on machinery, inputs, seed, labour, etc. An indirect measure of pesticide and fertilizer input.

19. *Tree density*: an important measure of the *dehesa* environmental health since both intensification and abandonment are reflected in the diminution of tree density. Besides, it determines the bird communities composition (open environment species vs forestry species).

Out of these 19 selected indicators 14 are applied to dry cereal farming and 12 to *dehesas* (Tables 10.2 and 10.3). The two agrosystems share seven indicators (37%) while seven and five indicators are specific to each agrosystem, respectively.

Eight dry cereal farming indicators and five *dehesa* indicators are highly relevant and have adequate statistics (type A, 68%), there being only three (23%) common to both agrosystems. Only one indicator for dry cereal farming was found to be moderately relevant with available statistics (type B, 5%). There were eight highly relevant indicators with no appropriate statistics available (type C, 42%), four (50%) being common to both agrosystems, one specific to dry cereal farming and three specific to *dehesas*.

Regarding the relationship between the indicators and the overall environmental value of the agrosystems, it should be noted that a total of four indicators in dry cereal farming and three in *dehesas* are hump-shaped.

The selected indicators must frequently address the environmental issues of diversity and soil conservation while pollution and fire prevention are less well represented in both agrosystems.

Finally, intensification processes seem easier to address with indicators than abandonment. For dry cereal farming, eight and two of the selected indicators analyse each process, respectively, and four indicators are valid for both trends. For the *dehesa*, there are six indicators associated with intensification, and only two of them are linked to abandonment. Four indicators address both processes.

PROBLEM ANALYSIS AND FUTURE NEEDS

The environmental systems pose serious difficulties as regards their assessment by indicators. The main limitations are a lack of linearity, immediacy and unequivocal causality in the response by the ecosystems to specific pressures. Obviously, this comment is also valid when using the indicators to assess policy (e.g. CAP) impacts on the environment. This is the reason why all of the indicators considered in this chapter can be catalogued as 'driving-force indicators' (OECD, 1997). The measurement of the environmental health of the system is indirect in that the indicators do not measure changes in the 'state' of the ecosystem (e.g. the number of great bustards or concentration of nitrogen in groundwater), but rather change in variables that affect the state and where the relationships are well understood, at least qualitatively. Hence, we are not proposing 'state indicators', trends in which cannot be attributed directly to the application of policy, and whose measurement would be extremely expensive. Moreover, such 'state variables' pose problems that invalidate their use as indicators: they are subject to a multitude of influences unrelated to the actual process to be measured, and may well lag behind the source of disturbance. For example, the presence of the great bustard may depend on factors unrelated to agricultural management or policy, such as the intrinsic population dynamics of the species or the existence or non-existence of disturbance due to human activity.

An exceedingly important aspect that guided the selection of the indicators was the availability of suitable data in official statistics. The problem of data suitability has two dimensions. First, the spatial scale of the data must be relevant for the process to be analysed. In this case, the scope should be smaller than NUTS 3, which is excessively heterogeneous, both environmentally and socially. Thus, two linked aspects must be considered: (i) the spatial continuity of the agrosystem, and (ii) homogeneity in the intensification–extensification–abandonment process. Obviously there is a problem related to the fact that these two aspects express themselves at all scales. The *comarca* (rural county) in Spain might be a good approach for dry cereal farming systems, which are relatively uniform within a single *comarca*, both in terms of the intensification–extensification–abandonment gradient and the spatial continuity of the agrosystem. However, this scale is less rated for the *dehesa*, a system less continuous and frequently interspersed with other agrosystems such as treeless grasslands and crops even within a single *comarca*. Furthermore, *dehesa comarcas* have farms with different degrees of intensification–extensification–abandonment, and this limits the use that can be made of average *comarca* data.[2]

Figure 10.3 exemplifies this scale-dependent suitability of data as a function of the spatial continuity of systems and processes. Here, three different *comarcas*, (a), (b) and (c), have different levels of internal continuity in systems and processes, depending on that of the farms present: some farms (in white) show an optimal system configuration (adequate tree density and stocking levels) and thus optimal environmental conditions; some others (shaded) offer intermediate conditions (because of overgrazing or crop extension resulting in reduced tree density); and the rest (in black) show the worst configuration (abandonment or irrigation conversion), yielding, as a consequence, the worst environmental conditions. The average values for each of the three *comarcas* (the statistical data at county level) are the same, while the actual configurations and environmental conditions remain quite different. The average at

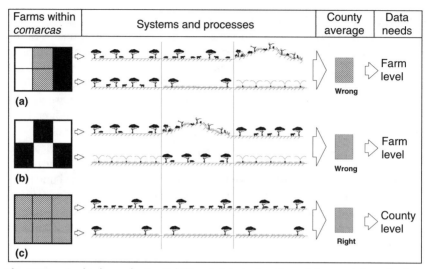

Fig. 10.3. Scale-dependent suitability of data in function of the spatial continuity of systems and processes.

county level is meaningless in cases (a) and (b) for which data at farm level are needed.

The second dimension concerns the extent to which the available statistics are suitable for application to the problem in hand. There is an absence of statistics for indicators that are relevant from an environmental perspective because they are directed at measuring farm production and are focused on a socio-economic analysis of the zone rather than an environmental analysis of the agrosystems. There are no official statistics on several particularly important indicators (e.g. density of trees in *dehesas* or differentiation between extensive and intensive livestock husbandry, meaning rough grazing and housed livestock systems, respectively).

The analysis of the two dimensions implies, first, that certain indicators need to be measured at a local and/or farm level, and, secondly, that official agricultural statistics should cover additional, more environment-related parameters if they are to be used to establish agri-environmental indicators.

Another facet arising from the results of this study is linked to the need to distinguish the indicators for each agrosystem, all the more so if the agrosystems differ in the relative importance of their livestock and crop uses. The higher importance of grazing in the *dehesa* system in comparison with dry cereal farming causes 45.5% of indicators of the system to be related to grazing in the former system and only 21.4% in the latter. We have not found a high degree of concordance between the indicators selected for two extensive systems in the Mediterranean area. What degree of concordance can be thus expected between intensive and extensive systems? Furthermore, a large number of the selected indicators do not provide an unambiguous measurement of the trends in the environmental value of the agrosystems. Instead, they reflect these trends along a gradient. For example, high livestock density in an extensive husbandry system reflects overgrazing problems and hence negative impacts on the environmental variables of the system. Low levels of the indicator, on the other hand, reflect conditions of undergrazing, which are negative for the system, either because of abandonment or transformation into crops (intensification). The definition of the threshold, beyond which the relationship changes sign, is absolutely specific to the individual conditions and characteristics of the agrosystem being analysed.

The next aspect that must be stressed is related to the need to use indicators that measure both intensification and abandonment. In extensive agrosystems, both processes caused by the application of the CAP are important from an environmental perspective. Thus, we see a third group of indicators that refer to the two types of processes. The first two cases are obvious. The 'area of irrigated land' indicator is an unequivocal measure of the degree of intensification in the two agrosystems analysed here, while 'area of scrub' measures the degree of abandonment of the production practices in both types of system. The 'extensive livestock husbandry' indicator serves to illustrate the third category. High values of the indicator reflect intensification by overgrazing and low values initially reflect abandonment.[3] Because the indicator operates in a hump-shaped way, its intermediate values are those that are unquestionably positive for the environmental health of the system. Finally, it must be noted that the two types of processes can coexist spatially within a single administrative area, even at a county scale, which once again highlights the inadequacy of the present statistics.

All of the above-mentioned aspects cause serious impediments to the selection and application of a single indicator system in all of the situations we wish to evaluate. It thus seems impossible to construct a general (common) system of agri-environmental indicators for the purpose of evaluating the impact of the CAP on the environment. Moreover, the indicator system cannot be made operative using the statistical databases currently available. The list of the indicators must therefore be peculiar to each agrosystem and the value of the indicators must be calculated at a county or a farm level in most cases.

However, a test on the application of the indicator system to two dry cereal *comarcas* (*c.* 150 000 ha farming counties) shows the potential value of the method even though official statistics cover nine indicators only. On the one hand, indicators clearly differentiate between *comarcas* with distinct natural values (Fig. 10.4): the most dense and stable great bustard population on the Iberian Peninsula is located in Campos-Pan *comarca* (Zamora Province), and the Peñaranda de Bracamonte comarca (Salamanca) has a much smaller population (Alonso and Alonso, 1990a). Thus, the value of all extensive-system indicators (area of fallow land, area of legume crops and degree of extensification) is higher in Campos-Pan, while the opposite is true with the indicator 'area of crops', which is four times higher in Peñaranda de Bracamonte. On the other hand, the temporal trend of indicators highlights an intensification

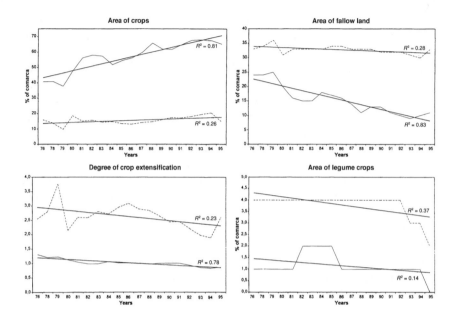

Fig. 10.4.   Time trends in values of four agri-environmental indicators in two dry cereal farming *comarcas* in Castilla y León (central Spain): Peñaranda de Bracamonte (solid line) and Campos-Pan (dashed line). See text for a detailed explanation of the indicators. The tendency line and the proportion of variability explained by it (R^2) is also included.

in the agrarian system in both *comarcas*, particularly so in the one with a lower natural value. These patterns are also evident in the unrepresented indicators (area of irrigated land, area of low scrub and barren land, area of dry pastures, area of scrub and diversity of agricultural substrata). They introduce small nuances such as a degree of abandonment in Campos-Pan (denoted by the increase in the area of scrub) and a small increase in the area of dry pastures and of low scrub and barren land in Peñaranda de Bracamonte.

CONCLUDING REMARKS

Several conclusions can be made from this analysis. First, the agri-environmental indicators in extensive systems should be applied at a county level, or even at a local or farm scale, depending on the spatial continuity of agrosystems and trends of intensification–extensification–abandonment. Otherwise, the resulting figures may not have a clear significance. This need for at least a county-based orientation has a similar precedent in the CAP reform of the cereal sector in 1992. Secondly, agricultural statistics available at a regional, national and EU scale do not provide information for extremely important indicators of the environmental value of these agrosystems. A system of agri-environmental indicators should therefore broaden these statistics so as to include the proposed indicators for the different agrosystems. Thirdly, many indicators are specific to each agrosystem. While some indicators have an equivalent application in the two types of system, others are applied in a different sense in each one, and some indicators are exclusive to one of the two agrosystems. This is corroborated by the fact that values with a universal application cannot be found in the hump-shaped operating indicators. Finally, for these extensive systems, indicators should not only cover intensification, but also abandonment, since both processes are important in terms of evaluating the environmental impact of the CAP. The relationship between each indicator and the type of process to which it refers is generally unequivocal, although in a few cases a single indicator makes reference to the two processes.

In order to overcome the limitations detected in the official statistics we propose a three-stage approach to evaluate the impact of the CAP on the environment:

1. Make an inventory of the European agrosystems.
2. Draw up a list of indicators that are relevant to each agrosystem, designating the direct negative, positive or hump-shaped mechanism to each one, along with the main driving force in play reflected by them. Some will be common to several agrosystems but even these will have specific attributes that force them to be considered as dependent on each system in each case.
3. Finally, a mechanism must be established for collection of information at NUTS 4 or farm level. For this purpose there appear to be four possibilities:

- To plan a Green Farm Survey specifically aimed at collating information to calculate the value of all of the defined indicators. This could be done for 5-year periods in the same way as the existing Farm Structure Survey (FSS).
- To extend the objectives of the present FSS in such a way that the questionnaire will request information needed for the indicators and to calculate their value every 5 years.
- To build a Green Farm Network specifically aimed at an annual collection of the information needed to calculate the value of the indicators on a yearly basis.
- To extend the objectives of the Farm Accountancy Data Network (FADN), including the information needed to calculate the value of the indicators every year.

The choice between these options will depend on the criteria used in the assessment which, in our opinion, should include at least the dimensions of cost-effectiveness and statistical significance. Our preliminary valuation suggests the Green Farm Network or the Green Farm Survey as the most cost-effective options.

Finally, we must consider the potential of using Geographical Information Systems technology in an attempt to overcome the limitations of the availability of statistical data. The fast-growing development of satellite imaginery will soon make detailed information on land use available, although information on several items (e.g. extensive husbandry management, fertilizer inputs) will always rely on statistics. When this arises, traditional data collection approaches within administrative boundaries might be complemented so as to become a reliable source of information for monitoring agroecosystem change.

NOTES

[1] For a synthesis see also De Juana (1988, 1989) and Suárez *et al.* (1996).

[2] The consideration of this feature of the *dehesas* implies that none of the highly relevant indicators for this agrosystem can be constructed or measured using suitable statistics. In Table 10.3 this has been noted with an A (C) under 'Type' heading.

[3] 'Initially' because these low values can also be related to livestock intensification, i.e. high values of the 'intensive livestock husbandry' (housed) indicator.

ACKNOWLEDGEMENTS

The present study has been carried out with financial support from the Commission of the European Communities, Agriculture and Fisheries (FAIR) specific RTD programme, CT95-0274, 'Implementation and Effectiveness of EU Agri-environmental Schemes established under Regulation 2078/92'. It does not necessarily reflect the programme's views and in no way anticipates the Commission's future policy in this area.

REFERENCES

Alonso, J.C. and Alonso, J.A. (eds) (1990a) *Parámetros Demográficos, Selección de Hábitat y Distribución de la Avutarda* (Otis tarda) *en Tres Regiones Españolas.* ICONA, Colección Técnica, Madrid.

Alonso, J.A. and Alonso, J.C. (1990b) *Distribución y Demografía de la Grulla* (Grus grus) *en España.* ICONA, Colección Técnica, Madrid.

Baldock, D., Beaufoy, G., Brouwer, F. and Godeschalk, F. (1996) *Farming at the Margins. Abandonment or Redemployement of Agricultural Land in Europe.* LEI-DLO, London and The Hague.

Beaufoy, G., Baldock, D. and Clark, J. (eds) (1994) *The Nature of Farming: Low Intensity Farming Systems in Nine European Countries.* Institute for European Environmental Policy, London.

Bernáldez, F.G., Rey Benayas, J.M., Levassor, C. and Peco, B. (1989) Landscape ecology of uncultivated lowlands in Central Spain. *Landscape Ecology* 2, 3–18.

Blanco, J.C. and González, J.L. (eds) (1992) *Libro Rojo de los Vertebrados de España*. ICONA, Colección Técnica, Madrid.

Campos, P. (1993) Valores comerciales y ambientales de las *dehesas* españolas. *Agricultura y Sociedad* 66, 9–41.

De Juana, E. (1988) Areas importantes para las aves esteparias. *La Garcilla* 71, 18–19.

De Juana, E. (1989) Las aves esteparias en España. In: *Seminario Sobre Zonas Áridas en España*. Real Academia de Ciencias Físicas, Exactas y Naturales, Madrid, pp. 199–221.

De Juana, E., Martín-Novella, C., Naveso, M.A., Pain, D. and Sears, J. (1993) Farming and birds in Spain: threats and opportunities for conservation. *RSPB Conservation Review* 7, 67–73.

Díaz, M., Campos, P. and Pulido, J. (1997) The Spanish *dehesas*: a diversity in land-use and wildlife. In: Pain, D. and Pienkowski, M. (eds) *Farming and Birds in Europe: The Common Agricultural Policy and its Implication for Bird Conservation*. Academic Press, London, pp. 178–209.

Domínguez Lozano, F., Galicia Herbada, D., Moreno Rivero, J., Moreno Saíz, J.C. and Sainz Ollero, H. (1996) Threatened plants in peninsular and balearic Spain. A report based on the EU habitats directive. *Biological Conservation* 76, 123–133.

Donázar, J.A. (1993) *Los buitres ibéricos*. J.M. Reyero, Madrid.

Florín, M., Besteiro, A.G. and Montes, C. (1996) Ecological support to model the hydrological function of Mediterranean wetlands: methodological management implications. Poster presented to INTECOL International Wetlands Conference, Perth, Australia.

Florín, M., Besteiro, A.G. and Montes, C. (1998) Ecological support to model the hydrological function of Mediterranean wetlands: methodological management implications. *International Journal of Ecology and Environmental Science*, in press.

Garza, V. and Suárez, F. (1990) Distribución, población y selección de habitat de la Alondra de Dupont, *Chersophilus duponti*, en la Península Ibérica. *Ardeola* 37, 3–12.

González, L.M., Bustamente, J. and Hiraldo, F. (1990) Factors influencing the present distribution of the Spanish Imperial Eagle *Aquila adalberti*. *Biological Conservation* 51, 311–319.

González Bernáldez, F. (1991) Ecological consequences of land abandonment in Central Spain. *Options Mediterranéennes* 15, 23–29.

McKenzie, D.H., Hyatt, D.E. and McDonald, V.J. (1992) *Ecological Indicators*, Vols 1 and 2. Chapman & Hall, London.

Marañón, T. (1985) Diversidad florística y heterogeneidad ambiental en una *dehesa* de Sierra Morena. *Anales de Edafología y Agrobiología* 44, 1183–1197.

Martínez, C. (1994) Habitat selection of the little bustard *Tetrax tetrax* in cultivated areas of Central Spain. *Biological Conservation* 67, 125–128.

Ministerio de Medio Ambiente (1996) *Indicadores Ambientales. Una Propuesta para España*. Secretaria General de Medio Ambiente, Madrid.

MOPTMA (Ministerio de Obras Públicas, Transporte y Medio Ambiente) (1995) *Estrategia Nacional para la Conservación y el Uso Sostenible de la Diversidad Biológica*. Secretaría de Estado de Medio Ambiente y Vivienda, MOPTMA, Madrid.

OECD (1997) *Environmental indicators for agriculture*. OECD, Paris.

Peco, B. (1989) Modelling Mediterranean pasture dynamics. *Vegetatio* 83, 269–276.

Rey Benayas, J.M., Bernáldez, F.G., Levassor, C. and Peco, B. (1990) Vegetation of aquifer discharges as conditioned by ground-water flows and solute transfer: the case of the Douro Basin, Central Spain. *Journal of Vegetation Science* 1, 461–466.

Suárez, F., Naveso, M.A. and De Juana, E. (1996) Farming in the drylands of Spain: birds of pseudosteppes. In: Pain, D. and Pienkowski, M. (eds) *Farming and Birds in Europe: The Common Agricultural Policy and its Implication for Bird Conservation.* Academic Press, London, pp. 297–330.

Suárez, F., Martínez, C., Herranz, J. and Yanes, M. (1997a) Conservation status and farmland requirements of Pin-tailed Sandgrouse *(Pterocles alchata)* and Black-bellied Sandgrouse (Pterocles orientalis) in Spain. *Biological Conservation* 82, 73–80.

Suárez, F., Oñate, J.J., Malo, J.E. and Peco, B. (1997b) La aplicación del Reglamento Agroambiental 2078/92/CEE y la conservación de la naturaleza en España. *Revista Española de Economía Agraria* 179(1), 267–296.

Tucker, G.M. and Heath, M.F. (eds) (1994) *Birds in Europe. Their Conservation Status.* BirdLife Conservation Series No. 3. BirdLife International, Cambridge.

Viedma, M.G. and Goméz-Bustillo, M.R. (1985) *Revisión del Libro Rojo de los Lepidópteros Españoles.* ICONA, Ministerio de Agricultura Pesca y Alimentación, Madrid.

Viejo, J.L., Viedma, M.G. and Martínez, E. (1989) The importance of woodlands in the conservation of butterflies (Lep. Papilionoidea and Hesperioidea) in the center of the Iberian Peninsula. *Biological Conservation* 48, 101–114.

Towards Environmental Pressure Indicators for Pesticide Impacts

11

Arie Oskam and Rob Vijftigschild

INTRODUCTION

There is a growing need to quantify the environmental impact of both broad policy areas and of specific inputs and outputs by means of indicators (e.g. Brouwer and Van Berkum, 1996; OECD, 1997). An indicator summarizes a set of related information in one single variable, which is presented in a recognizable form and a workable dimension. This observation holds specifically for pesticides, which form a heterogeneous group of products designed to combat several thousands of different weeds, diseases and pests, and which consist of several hundreds of different active substances (Pease *et al.*, 1996).

Because of the dangerous aspects of applying pesticides, an extensive regulation mechanism has been developed. Here, the authorization of pesticides is the most important element. Together with the authorization, the definition of proper use (crops, quantities, circumstances, etc.) to be used on the labels of pesticide containers or packages forms the most important element in the control of pesticide application. A second line of development is to enhance the information available to farmers and contractors about the consequences of applying pesticides and the possible ways in which harmful impacts could be reduced. Within this second approach, long-term developments have been stimulated mostly by technological changes: integrated pest management or integrated crop management (Wijnands and Vereijken, 1992). More fundamental changes derive from the adoption of ecological/organic farming methods.

Besides regulation, other instruments have been used. Within the European Union (EU) Sweden, Denmark, The Netherlands and Finland operate plans to reduce the risk and application of pesticides in successive steps with clearly defined targets for volume, emission or number of treatments (Oskam *et al.*, 1998). Germany has relatively strict rules on the authorization of pesticides, together with compensation for farmers in water-protection zones and other environmentally sensitive areas (Heinz *et al.*, 1995). Levies are used in Denmark, Sweden and, to a lesser extent, in Finland. Often, combinations of instruments are applied and this is most clear in plans to reduce the use and risk of pesticides.

No matter what instruments are chosen, good indicators are essential for the provision of the right information and incentives. If good indicators are not available, it will be difficult to assess developments and the effects of policy instruments in the highly intricate area of pesticides and their effects on environment and food safety.

This chapter starts with the definition of an environmental indicator and the methodology involved in the development of indicators and gives examples derived from the literature. In the next section a number of environmental risk indicators are proposed, the authors' contribution comprising the definition of a number of operational indicators, each reflecting a particular aspect of the environmental burden of pesticides. Council Directive 91/414 of the EU plays an important role in present policy and this will change the registration of pesticides in member countries substantially. However, the presence and quantities of pesticides and residuals in the air, soil, groundwater and surface water represent the environmental impact of pesticides. The next section is a general approach to the selective reduction in the number of environmental indicators. Such an approach can be used if sufficient background information is available. The chapter ends with a discussion and conclusions.

DEFINITION OF AN ENVIRONMENTAL INDICATOR AND THE APPLICATION OF THIS DEFINITION TO PESTICIDES

OECD Approach

OECD (1994) starts with the following description of an agri-environmental indicator:

> The main objective [of an agri-environmental indicator] is to assist policy makers in their evaluation of current and alternative agricultural and environmental policy measures, by quantifying the relationships between agricultural activities and the environment. Such indicators would help to provide information on the positive and negative impacts of agriculture on the environment, and the effects of changes in the environment on agriculture.

This type of approach has been developed at the request of the 1989 G-7 Economic Summit at Paris, where the OECD was invited to develop environmental indicators to support policy-makers.

A number of observations can be made:

1. An indicator is defined as the *relationship* between agricultural activities and the environment, which should be quantified. There seems to be no 'objective' state of the environment.

2. The existence of an environmental indicator depends on the policy area; indicators exist in relation to policies.

3. It seems sufficient that an indicator provides information on the positive and negative impacts of agriculture on the environment; here an ordinal scale would be adequate. But, as indicated above, the relationship should also be quantified, which gives rise to evaluation at a cardinal scale.

4. There is a specific emphasis on how agriculture influences the environment.

The development of environmental indicators by the OECD has been carried out within the pressure–state–response (PSR) framework: a policy-orientated approach.[1] Here the most important relationship is between the human activities causing the pressure and the state of the environment. The second relationship refers to the response or policy reaction on the change in the state of the environment. The PSR/DSR framework is, in our opinion, much broader than an environmental indicator, which refers to the 'state of the environment'. Due to its strong focus on policy issues, the conceptual framework of the OECD may provide limited indications of the state of the environment. It would be better to use the names 'agri-environmental pressure indicator' for the first relation within the PSR framework and 'policy/societal reaction indicator' for the second relation (between the state of the environment and the societal response).

It will be clear that the first relation is easier to quantify than the second. Several models have been developed to simulate pollution from agricultural production, but empirical tests of these models are often lacking (e.g. Mulkey *et al.*, 1986). A preliminary investigation was carried out for The Netherlands by Koops *et al.* (1996) on the relation between pesticide application and the appearance and behaviour of pesticides in the soil. The measured amounts of seven pesticides corresponded reasonably well with the amounts calculated in the simulation model, PESTLA. Hoekstra (1996) reported on the strategy of RIVM (National Institute of Public Health and Environmental Protection) regarding the relation between policy measures and environmental pressure in the area of pesticides, but this still differs from a 'societal response'.

Alternative Approach

In this chapter, we start from a different definition of an environmental indicator. This definition is, on the one hand, much less ambitious, because it is focused only on the 'state of the environment',[2] but on the other hand, the definition tries to be more precise in defining how the state of the environment is quantified. It follows more closely the tradition of environmental indices (e.g. Ott, 1978; Hammond *et al.*, 1995).

We use the following *definition* of an environmental indicator: an indicator is a one-dimensional representation of a set of information with respect to the state of the environment or a particular aspect of the state of the environment, presented on a cardinal scale. It should be immediately clear from the indicator, whether the state of the environment improves or deteriorates. The quality of the indicator is better as long as an improvement of the indicator represents an improvement of the state of the environment, according to both scientific standards and to the perception of the general public. In practice, indicators might not fulfil all the requirements of this definition, because they should be operational and not too costly.

It is assumed that the set of information consists of a number of *basic variables*. These basic variables represent relevant aspects of the state of the environment. The importance of a basic variable is represented by its weight, which may depend on the level of the basic variable.

The weighted indicator, which summarizes either flows or stocks (or even a combination of both), can be defined as follows:

$$EI = \sum_{n=1}^{N} w_n(V_n) \times V_n \qquad [11.1]$$

where:

   EI = environmental indicator/index;
   w = weight, which depends on the type and level of the basic variable V;
   V = basic environmental variable/indicator;
   n = counter of the number of basic variables (total N, which depends on the particular indicator).

Clearly, the most important issue here will be to find a selection of basic variables together with their relative weights. Different weights at different levels of the basic environmental variable may arise if, for example, thresholds are relevant. Such weights are related mainly to the environmental damage caused by different levels of basic environmental variables. Although this is a highly intricate area of research, it is possible to present some examples for illustrative purposes.

Figure 11.1 illustrates a number of different situations. If relation (a) is relevant, then a constant w is correct. In situation (b), however, a constant w applies after the threshold level t_b. Situation (c) suggests two levels of w, switching at t_c and redefining the basic environmental variable to a two-level variable. If situation (d) applies, w increases with the level of the basic variable. Ott (1978, Chapter 2) deals more generally with these issues.

Clearly it will be a simplification if w is independent of the level of the basic environmental variable V – as has been illustrated in situation (a) of Fig. 11.1, or if the EI depends on only one basic environmental variable ($K = 1$). Later in this chapter we consider the reduction in the number of variables that play a role in composing an indicator. The weighing of different environmental

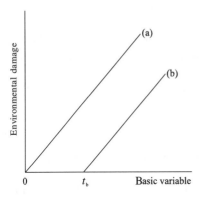 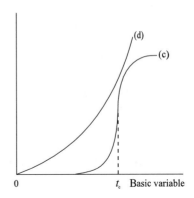

Fig. 11.1. Relations between a basic environmental variable and the environmental damage caused by pesticides.

indicators will not be dealt with explicitly: ultimately this will be a matter of political and scientific consensus.

Some Examples in the Literature of Environmental Indicators in Relation to Pesticides

First we provide two examples of flow indicators. The best-known flow indicator is the 'intensity of use' indicator, in which annual sales of pesticides are related to the area of production. This indicator has been used in several studies to indicate the pressure of pesticide use in agriculture on the environment (e.g. Brouwer *et al.*, 1994). The numerator can be specified by product group (herbicides, etc.) or active substance (e.g. lindane, atrazine), but is usually expressed in kilograms of active substance (a.s.). The denominator can be specified by agricultural sector (e.g. arable farming) or crop (e.g. wheat; Landell Mills, 1996), but is most often expressed in hectares of arable land and land under permanent crops. The specification depends on the problem to be studied and the statistics available.

Another example of a flow indicator, which goes one step further, is environmental exposure to pesticides (EEP), in which the intensity of use is multiplied with a dilution/degradation factor, which varies for the components air, soil and water (Wijnands and Van Dongen, 1995). This indicator has been used in research on integrated farming systems. However, this is only one example of the calculation of predicted environmental concentration (PEC) resulting from pesticide use in agriculture. An overview can be found in Beinat and Van den Berg (1996). Both flow indicators will be developed in more detail later in this chapter.

Secondly, we provide two examples of a stock indicator. The best-known stock indicator is 'monitoring results of the presence of pesticides in surface water'. This indicator has been used to indicate the hazard of pesticides in surface water and will be further elucidated later in the chapter. Normally, results are provided for certain active substances (Faasen, 1994; SSLRC, 1996). Another example of a stock indicator is 'monitoring results of the presence of pesticides in soil'. As far as we know, the availability of monitoring results is extremely limited. Those monitoring results of the few pesticides that are available, serve to validate simulation models (Koops *et al.*, 1996).

ENVIRONMENTAL INDICATORS IN RELATION TO PESTICIDE USE[3]

To develop a number of workable indicators, both risk aspects and the quantity of pesticides should play a role. If no pesticides are used, there will be no direct environmental risk in relation to pesticides, which does not imply that the general state of the environment will improve, because risk of crop losses and greater land use for farming may work in the opposite direction. If low-risk pesticides are used (due to their intrinsic nature or to application method, etc.), then the environmental burden is also reduced.

To provide a more systematic overview, risk[4] is distinguished in relation to (i) the authorization of pesticides; (ii) drinking-water quality; (iii) surface-water quality; and (iv) more general overview indicators at farm level. The indicators are all *ex post* indicators, but they might use information that is developed specifically for *ex ante* judgement. If the same indicators are compared for two periods of time or for different (regions of) member states, they should indicate the (changing) level of risk. If appropriate, special attention is given to the weighting of different environmental risks. For all indicators, attention must be given to the availability of data, and in many cases only monitoring results with differences in strategies and detection methods are available.

Authorization

Risk aspects of pesticides, including environmental risk, are mainly managed by authorization: an *ex ante* performance check. In the EU, authorization has to be done according to 'uniform principles' (CEC, 1991, 1994). The uniform principles do not only focus on environmental risk (the impact on non-target species, such as birds, bees, other beneficial arthropods, aquatic organisms, earthworms and soil microorganisms), but also on efficacy, and the impact on plants, vertebrates, workers and on consumers. Weighting of all these aspects is done with specific decision-making principles. In principle, pesticides have to fulfil all the criteria to become authorized.

If the authorization of pesticides, together with the prescription and monitoring of proper use, is sufficient, then an adequate risk indicator would be based on the authorization. Here we define the lowest risk level ($R = 0$) and the highest risk level ($R = 1$). The following definition provides a risk indicator:

$$R = 1 - \left(\frac{1}{K} \sum_{k=1}^{K} \delta_k \right) \qquad [11.2]$$

where $\delta_k = 1$ if k is authorized according to the uniform principles and $\delta_k = 0$, if otherwise; K = total number of pesticides.

This would imply that there is no risk ($R = 0$): an acceptable risk, if all pesticides have been authorized under the uniform principles. A slightly more relevant measure would include the share of the particular pesticide ($k = 1, \ldots, K$) in the total application of pesticides, measured in kilogram active substance (s_k), which results in:

$$R_1 = 1 - \left(\frac{1}{K} \sum_{k=1}^{K} s_k \delta_k \right). \qquad [11.3]$$

It will be quite clear that the authorization of pesticides, which are already on the market (K does not change), according to the harmonized criteria, contributes to risk reduction and even more if those pesticides take a substantial share in the total application. Of course, the share of a pesticide which is not yet marketed is always zero. So, risk reduction will affect equation 11.3 only after the introduction of a new pesticide.

R_1 requires knowledge of at least the shares of authorized pesticides in total use, preferably in a recent year to keep the measure up-to-date. R needs less information and could be preferred for this reason. During July 1993 a total of 840 pesticides were on the market of the Member States (Oskam *et al.*, 1998, p.48). If one realizes that up to now only a few pesticides are officially authorized according to the uniform principles (listed on Annex I of Directive 91/414), it is clear that, at this moment, $R = 1$.

Risk Indicators in Relation to Intake of Drinking Water

Monitoring data on (ground- and surface) water intended for human consumption could indicate the risk of emitted pesticides to the environment. Such results could be presented as the number of samples that meet the relevant requirements of Council Directive 80/778 (the maximum permissible concentration for each active substance is $0.1\,\mu g\,l^{-1}$; CEC, 1980) in relation to the total number of samples for active substance:

$$R_{2,k} = 1 - (\text{samples}_{80/778/\text{EEC},k} \,/\, \text{samples}_{\text{TOTAL},k}). \qquad [11.4]$$

If all samples fulfil the requirements of Council Directive 80/778, $R_{2,k} = 0$ and no risk is perceived for drinking water. Observe that this indicator does not differentiate between the degree to which the threshold is exceeded. It focuses on a part of the risks of pesticides in a zero–one framework. This indicator neglects toxicological properties of the active substance – no weighting is involved. The only information that counts is whether the levels monitored, with a constant sampling and detection procedure, do not pass the $0.1 \, \mu g \, l^{-1}$ limit.

Within a more refined procedure, the level above the limit is included in the analysis. Here the following indicator is proposed:

$$\overline{R}_{2,k} = 1 - \left(\frac{1}{N_k} \sum_{n=1}^{N_k} \frac{q_{k,n}}{\overline{q}_k} \delta_n \right) \qquad [11.5]$$

where $q_{k,n}$ is the sample value of observation n $(n = 1, \ldots, N_k)$; $\overline{q}_{k,n}$ is the average sample value of the total number of observations N_k; δ_n is equal to 1 if the observation is above the limit and otherwise zero.

It will be clear that this method gives a large number (K) of risk indicators, between zero and one. This information could be aggregated into one indicator according to the following formula:

$$\overline{R}_2 = \sum_{k=1}^{K} sd_k \, \overline{R}_{2,k}$$
$$sd_k = \frac{\overline{q}_k}{\overline{q}}. \qquad [11.6]$$

Here $\overline{q}$ is the average load of all observations.

Data are available although still not from an ideal monitoring programme. In Article 12 of Directive 80/778 it is laid down that member states shall take all necessary steps to ensure regular monitoring of the quality of water intended for human consumption. Heinz *et al.* (1995) described the monitoring systems, from which some results in Table 11.3 have been derived. SSLRC (1996) provided a summary. The question is whether a selection of important active substances can or should be made. For atrazine, considerable information is available (Bergman and Pugh, 1994).

Risk Indicators in Relation to Surface-water Quality

Monitoring surface-water quality could indicate the risk of emitted pesticides to the environment (the impact on non-target aquatic organisms). Monitoring data could be presented as the number of samples in which a certain pesticide is detected in relation to the total number of samples. The monitoring results could also be presented as the number of samples that meet the relevant requirements of Council Directive 91/414 (the maximum toxicity exposure ratio) in relation to the total number of samples for active substance k.[5] This means that in this indicator weighting is involved. The proposed indicator is:

$$R_{3,k} = 1 - (\text{samples}_{91/414/\text{EEC},k} \, / \, \text{samples}_{\text{TOTAL},k}) \qquad [11.7]$$

The indicator focuses on a part of the environmental risks of pesticides, and uses toxicological properties of the active substance (at single-species level) to weight the risk of each pesticide according to its expected impact. Further refinements or aggregations can be established similar to the risk measure for drinking water.

Risk Indicators in Relation to Agricultural Use of Pesticides

Monitoring results of pesticide use or active substances (a.s.) could indicate the risk of emitted pesticides to the environment. The amount used can be related to the total land area, the area of cultivated land or the area of certain crops (j) (which are all common statistics, at least at country level). The proposed measure is:

$$RM_{1,j} = \text{use of pesticides or a.s. for crop}_j \, / \, \text{area of crop}_j \qquad [11.8]$$

(The initials RM are used for risk management indicators; these are all positive or zero with no upper limit.)

This is a rough indicator of risk. If, with respect to the risk for the environment, it is assumed that all used pesticides are emitted, this is a kind of worst case. In this measure no toxicity is involved. Much information is available, especially for the total of a.s.

An alternative would be to relate pesticide use to crop yields (in kg) (or crop values in constant prices). This would at least correct for large differences in yields (or crop quality):

$$RM_{2,j} = \text{use of pesticides or a.s. for crop}_j \, / \, \text{crop yield (in kg)} \qquad [11.9]$$

With this indicator, information on the maximum emission of pesticides per hectare is lost.

The risk of emitted pesticides to the environment, and their impact, could be measured by extending equation 11.8 or 11.9 with environmental factors. Use, which in this case should be known in active substance (k) per ha of crop (j), is multiplied by the corresponding 'environmental impact quotient' (EIQ) or with 'environmental impact points' (EIP). The EIQ was first described by Kovach *et al.* (1992) and critically discussed by Dushoff *et al.* (1994) and Pease *et al.* (1996). Variants of the EIQ are 'the EIPs of the environmental yardsticks', developed by Reus and Pak (1993). Both systems are developed as an aid in IPM. The proposed indicator on the basis of equation 11.9 is:

$$RM_{3,j} = \sum_{k=1}^{K} \left(\text{use of a.s.}_k \times EIQ_k \right) \text{for crop}_j \, / \, \text{area of crop}_j \qquad [11.10]$$

or

$$RM_{4,j} = \sum_{k=1}^{K} \left(\text{use of a.s.}_k \times EIP_k \right) \text{for crop}_j \, / \, \text{area of crop}_j \qquad [11.11]$$

With these indicators, exposure as well as toxicity (of a single species) is involved. To make them suitable as an aid in IPM, several assumptions have to be made and, in the case of the EIQ, this includes the weighting of environmental aspects. For this reason, one should be careful when applying these indicators for purposes other than those for which they were developed. Similar results would be obtained by applying the measures EIQ or EIP to equation 11.9, resulting in risk management measures that focus on a per unit of product basis. Another approach is based on the number of applications per hectare, which is seen as an indicator of the impact of pesticide use on biodiversity.

Information is not available for several EU member states. Pease *et al.* (1996) developed a methodology to rank pesticides by hazard (including human health). In this methodology weighting is done as explicitly as possible. Table 11.2 shows a numerical overview, based on Californian pesticide use data.

It is still not clear which indicator provides most information on the various aspects of risk that play a role in the use of pesticides. Clearly, the measures defined in equations 11.10 and 11.11 require and provide more detailed information. As always in reducing the dimension from many to one, information is lost.

SELECTING INDICATORS ON THE BASIS OF PRINCIPAL COMPONENTS ANALYSIS

Several environmental indicators have been proposed. Equation 11.1 provides a method for combining this information. Here weights should be provided, but they are often not available in research nor is it usually known whether the indicators provide independent information. This produces further difficulty since weighting several possibly dependent indicators is more difficult than weighting a few independent ones.

Two principles are important:

1. To select the minimum number of independent environmental indicators, which provide a good description of the 'state of the environment' and/or the flow of emissions to the environment.
2. To see which observable indicators function best in representing the flow to the environment and/or the state of the environment.

One method for combining this information is principal components analysis (PCA) (Theil, 1971). This method first selects a number of independent '*constructed*' environmental indicators or basic environmental variables, which reflect the information included in the set of indicators/variables.[6] The second step is to determine which *observable* variables function best in reflecting this information. To show the potential of PCA we provide three different examples, but the lack of a broad data set on pesticides in the EU in relation to the environment prevents an adequate illustration on that subject.

A Limited Set of Environmental Performance Data from EU Countries

We start with a small set of data on population, minerals, pesticides and the reduction of biodiversity (Table 11.1). This example is for illustrative purposes only.

The results included in Table 11.2 allow the following conclusions:

Table 11.1. Some environment-related variables for the countries of the European Union (source: EUROSTAT, 1995; Oskam *et al.*, 1998).

Country	Populat- ion km^{-2}	Nitrogen ha^{-1}	Phos- phate ha^{-1}	Endangered amphibian species	Endangered vascular plant species	a.s. ha^{-1}	kg a.s. per 1000 ECU crop
Austria	94	87	47	90	30	4.3	2.78
Belgium	330	221	79	100	24	13.8	3.98
Denmark	120	144	30	29	10	1.7	2.01
Finland	15	66	30	20	7	1.3	0.95
France	105	134	65	50	3	5.8	4.44
Germany	227	143	43	58	24	2.5	2.41
Greece	78	102	45	6	2	4.1	1.51
Ireland	51	373	147	33	20	12.5	4.00
Italy	189	76	55	24	12	6.5	2.78
Luxembourg	152	221[a]	79[a]	87	13	4.4	7.22
Netherlands	370	406	78	63	35	14.0	1.69
Portugal	107	43	24	8[b]	4	5.7	7.29
Spain	77	50	25	8	6	3.1	3.25
Sweden	19	63	16	46	9	1.1	1.99
United Kingdom	238	204	51	33	10	5.4	4.35

[a] Set equal to the observation for Belgium.
[b] Set equal to the observation for Spain.

Table 11.2. Results of a principal components analysis of some environmental performance data.

Variable	Factor 1 'Intensity'	Factor 2 'Biodiversity'	Factor 3 'Inefficiency'
Phosphate ha^{-1}	0.92	0.12	0.16
Nitrogen ha^{-1}	0.88	0.34	−0.06
Pesticides (kg a.s. ha^{-1})	0.84	0.36	0.07
Endangered amphibian species	0.11	0.88	0.20
Endangered vascular plant species	0.38	0.81	−0.27
Population density	0.33	0.74	−0.01
Pesticides (kg a.s. 1000 ECU^{-1} crop production)	0.11	0.02	0.98

- The first factor represents the intensity of arable land use as indicated by the applications of minerals and pesticides. There is a low weighting for population density and endangered vascular plant species.
- The second factor is related to biodiversity and population pressure in EU countries. There is still a positive relationship with the typical intensity variables.
- The third factor clearly illustrates that efficiency/inefficiency of pesticide use is independent of intensity and population pressure and represents a specific element in the analysis.

A quick scan of this set of (mainly) environmental variables is obtained by 'phosphate per hectare arable land', 'endangered amphibian species' and 'kilogram active substance of pesticides per unit crop production'. If it is difficult to select one of these variables, the method gives an indication of which variables can be used as alternatives.

Pesticide Use, Observed Emissions and Residues in Fruits and Vegetables

The second example is based on the observations of 'general' pesticide use, pesticide emissions and pesticide residues in fruits and vegetables (Table 11.3). Observations were available for six countries only. Although for practical purposes this data set is very limited, it is interesting to see which easy-to-observe variables best reflect the emissions and the residuals.

Table 11.4 allows the following conclusions:

Table 11.3. Impacts from pesticide use in several EU Member States.

Country	Percentage of samples of vegetables and fruits with residues[a]		Percentage of pesticides detected in ground and surfacewater[b]	
	Below norm	Above norm	Below norm	Above norm
Austria	?	?	0.9	9.1
Denmark	11.7	0.5	1.7	3.5
France	?	?	7.0	6.1
Germany	39.9	0.7	11.3	17.8
Greece	11.3	7.7	2.2	6.1
Italy	?	?	3.5	19.6
Netherlands	36.1	2.0	10.9	24.8
Spain	35.0	3.6	8.7	13.5
Sweden	9.4	0.4	?	?
United Kingdom	41.0	1.7	5.7	25.7

Sources:
[a] Van Klaveren (1997, p.39), most recent data for each country.
[b] Heinz *et al.* (1995, Annex 5), total of 230 pesticides.

Table 11.4. Results of a principal components analysis on six variables in relation to pesticides (six countries).

Variable	Factor 1 'Intensity'	Factor 2 'Efficiency'
Pesticides (kg a.s. ha⁻¹)	0.88	−0.30
Pesticides in ground- and surface water (above norm)	0.82	0.49
Pesticides in ground- and surface water (below norm)	0.81	0.33
Pesticides (kg a.s. 1000 ECU⁻¹ crop production)	0.05	0.90
% Samples of fruits and vegetables with residues (below norm)	0.68	0.71
% Samples of fruits and vegetables with residues (above norm)	−0.18	−0.62

- Pesticides measured in kg a.s. per hectare (intensity) is a good indicator of the pesticide level observed in groundwater and surface water, whether the observations are above or below the norm.
- Intensity of pesticide use says nothing about the residues in fruits and vegetables above the norm (correlation is negative).
- There is a positive relationship between the intensity of pesticide use and residues below the norm. This seems a plausible conclusion: if pesticides are used intensively they can be observed in the products, but that does not imply that these observations are above the norm.
- The efficiency of pesticide use is positively related to the level of pesticides (below the norm) in fruits and vegetables.[7]
- There is a clear negative relation between efficiency and the percentage of fruits and vegetables above the norm (*c.* 38% of the variance is related to the efficiency factor).
- A high efficiency is not strongly related to the level of emissions.

If the data are representative and these conclusions hold, then pesticide intensity is a good indicator of emissions but not of residues. Pesticide efficiency (as a second indicator) is related to residue levels, but the direction requires further investigation.

Environmental Performance Indicators and Pesticide Use in California

The third example focuses totally on pesticides, but the data relate to California (Pease *et al.*, 1996), see Table 11.5. This data set for 28 different pesticides is concerned with observable objective data and with three different subjective variables: linear rank, elicited rank and step rank.

Table 11.6 shows:

Table 11.5. Some environmental and health related variables for pesticides applied in California (source: Pease *et al.*, 1996, Tables A1 and A2).

Pesticide	Selection[a]	Oral LD50[b]	Acute illnesses	Avian LD50[b]	Fish LC50[b]	Well detections	Field half-life	Step rank[b,c]	Linear rank[b,c]	Elicited rank[b,c]
2,4-D	2	699	0	200	0.9	13	10	5	41	36
Acephate	1	980	0	140	1000	0	3	25	18	125
Aldicarb	2	1	75	1		34	30	2	28	3
Atrazine	2	1 780	0	2000	4.3	218	60	14	16	28
Benomyl	2	10 000	0.29	100	0.02	2	67	13	139	39
Captan	2	9 000	0	2000	0.02	3	2.5	36	142	131
Carbaryl	1	264	1.29	2000	1.95	2	10	34	51	129
Carbofuran	2	11	9.86	0.4	0.14	1	50	3	31	4
Chlorothalonil	2	10 000	0	4640	0.05	1	30	28	146	59
Chlorpyrifos	2	183	42.43	17.7	2.4	3	30	21	17	8
Cyanazine	1	288	11	400	10	0	14	49	8	19
Diazinon	1	350	5.71	3.2	0.09	7	40	8	19	14
Dicofol	1	890	9.57	265	0.11	0	45	16	20	16
Dimethoate	2	255	0.29	6.6	6	3	7	6	26	18
Diuron	2	3 400	14.71	5000	4.3	223	90	48	36	64

Hexazinone	1	1 690	1.71	2258	274	0	90	64	107	97
Phosmet	1	231	0.14	2009	0.07	0	19	40	39	24
Iprodione	1	4 400	1.71	930	2.25	0	14	132	126	117
Malathion	2	1 900	0	167	40	1	1	47	38	55
Mancozeb	1	11 200	0.14	1500	0.46	0	70	58	145	41
Maneb	2	7 990	1.14	1467	1.8	0	70	30	135	35
Metam sodium	2	1 909	0	500	0.08	0	7	19	2	30
Methamidophos	1	20	1.29	25	46	0	6	22	5	17
Methomyl	1	20	0.14	24.2	0.5	1	30	1	22	1
Mevinphos	1	7	0.14	1.34	0.01	0	3	4	9	5
Parathion	2	4	0.29	1.3	0.32	0	14	11	25	2
Simazine	2	5 000	0.57	5000	85	497	60	9	11	32
Trifluralin	1	10 000	1.57	2000	0.01	2	60	7	131	21

[a] Those pesticides marked with 2 were selected for a second calculation.
[b] A low value means a high toxicity, a high hazard.
[c] Ranks were calculated from 12 pesticide impact attributes, including the six mentioned in this table.

- The linear rank and the elicited rank largely determine the first two independent factors. This can be explained readily since they are based on 12 pesticide impact attributes. These two 'synthetic variables' make the third one, 'step rank', virtually superfluous.
- The third factor is well represented by 'well detections'.
- If the 'rank variables' were not available, the 'oral LD_{50}' could represent the first factor. For the second factor it is more difficult to find a directly observable indicator.
- Characterizing the factors is not easy: the first could be related to persistence, the second more to toxicity and the third to appearance in the groundwater.

To see whether the results were stable, a PCA was made of a selection of 15 pesticides (Table 11.7). The selection was quite arbitrary, but included insecticides, fungicides and herbicides as well as nematicides. The results show some differences compared with Table 11.6. Now all rank variables appeared in the

Table 11.6. Results of a principal components analysis on nine variables in relation to pesticides (based on 28 pesticides; see Table 11.5).

Variable	Factor 1	Factor 2	Factor 3
Linear rank	0.92	0.32	−0.02
Oral LD_{50}	0.85	0.19	0.24
Elicited rank	0.14	0.92	0.04
Fish LC_{50}	−0.44	0.64	0.01
Step rank	0.33	0.64	−0.06
Acute illnesses	−0.16	−0.45	−0.04
Well detections	−0.21	−0.05	0.92
Avian LD_{50}	0.29	0.26	0.83
Field half-life	0.45	−0.17	0.63

Table 11.7. Results of a principal components analysis on nine variables in relation to pesticides (15 pesticides selected from Table 11.5).

Variable	Factor 1	Factor 2	Factor 3
Elicited rank	0.91	0.04	−0.04
Oral LD_{50}	0.81	−0.13	0.45
Linear rank	0.74	−0.48	0.34
Step rank	0.71	0.08	0.02
Acute illnesses	−0.59	−0.24	0.25
Well detections	−0.05	0.90	0.39
Fish LC_{50}	0.05	0.89	−0.03
Avian LD_{50}	0.46	0.58	0.52
Field half-life	−0.04	0.19	0.91

first factor. 'Elicited rank' and 'well detections' were still good indicators, but 'field half-life' could be identified as the third indicator.

We concluded from the three examples that PCA provides a useful method for deriving a set of independent environmental indicators. If particular variables are difficult to measure or costly to obtain, alternatives might be identified. The more factors there are identified in the PCA, the greater the dimensionality of the environmental problem. Weighting each of these factors (or representing variables) will be necessary if a single environmental index is required.

DISCUSSION AND CONCLUSIONS

Environmental aspects of pesticides are so diverse that simple indicators are often insufficient. This leads to a situation where several (potentially) measurable indicators are considered. Principal components analysis forms a useful method to reduce a 'basket of indicators' to a small set of readily available indicators. If information on weights is available, then a weighted indicator (or environmental index) can be derived from several independent basic indicators/variables.

Because weights on relevant indicators change over time or differ between units being compared, there will be no unique indicator (or set of indicators) which will always be relevant. Good indicator selection, however, can prevent use of inadequate or insufficient indicators. In practice, the availability of homogeneous information is often the main bottleneck to progress.

The methodology outlined here only provides a representation of the information contained in the data set. This could be information at country, regional or local level, or information over a period of time (or various combinations). The set of variables included in the analysis, however, determines the relevance of the resulting (set of) indicator(s). It is important to note that the PCA methodology is based on linear relations between variables. If highly non-linear relations are present, the methodology may not work well.

Several recent approaches of the OECD (1994, 1996, 1997), although conceptually very elegant, lack the empirical basis for defining and selecting environmental indicators. Here a broad database is helpful. With respect to pesticides, the work of Pease *et al.* (1996) provides a good starting point. The dataset available for California can be useful in developing relevant environmental indicators for pesticides in the EU, at least for more intensive production regions.

NOTES

[1] Later the OECD shifted to the driving force–state–response (DSR) approach developed by the United Nations (OECD, 1997). Driving forces provide a wider background.

[2] A similar definition refers to flow variables reflecting the environmental burden or pressure of the environment. Indicators will be defined both in relation to flow and state variables.

[3] This section draws on Oskam *et al.* (1998, Appendix I) and provides further elaboration of OECD (1996).

[4] Here risk is defined as the probability of an event times the loss (or environmental damage) caused by that event. Pesticides cause a large range of 'events'. Often, most attention is given to determining the loss (or environmental damage).

[5] In this case, the toxicity in relation to the *measured* concentration and not the *expected* concentration, for each active substance is more than 100 for fish and *Daphnia*, and more than 10 for algae (derived from CEC, 1994).

[6] Here a set of n observations on m different variables is represented by the $(n \times m)$ matrix, X. The correlation matrix (R) of the m different variables is represented by a small number of orthogonal eigenvectors. This representation is best as long as the eigenvectors with the largest eigenvalues are selected. The 'binding coefficients' of the original variables to the eigenvectors determine the 'principal component'. Binding coefficients have been provided in Table 11.2. Often – and also in our example – a varimax rotation is applied to select a small number of principal components, which are clearly related to (or independent of) the original variables (Harman, 1960).

[7] About 50% of the variance in observations above the norm is related to the efficiency factor: $0.71^2 \times 100 = 50\%$.

REFERENCES

Beinat, E. and Berg, R. van den (eds) (1996) *EUPHIDS, a Decision Support System for the Admission of Pesticides.* RIVM Report No. 712405002, National Institute of Public Health and Environmental Protection (RIVM) RIVM/VU/FhG/ICPS, Bilthoven.

Bergman, L. and Pugh, D.M. (eds) (1994) *Environmental Toxicology, Economics and Institutions: the Atrazine Case Study, Economy and Environment,* Vol. 8. Kluwer Academic, Dordrecht.

Brouwer, F.M and Berkum, S. van (1996) *CAP and Environment in the European Union. Analysis of the Effects of the CAP on the Environmental Assessment of Existing Environmental Conditions in Policy.* Wageningen Pers, Wageningen.

Brouwer, F.M., Terluin, I.J. and Godeschalk, F.E. (1994) *Pesticides in the EC (PES-A 1/1), LEI Onderzoeksverslag 121.* Agricultural Economics Research Institute, 's Gravenhage.

CEC (1980) Council Directive of 15 July relating to the quality of water intended for human consumption (80/778/EEC). *Official Journal of the European Communities* L 229.

CEC (1991) Council Directive of 15 July 1991 concerning the placing of plant protection products on the market (91/414/EEC). *Official Journal of the European Communities* L 230.

CEC (1994) Council Directive of 27 July establishing Annex VI to Directive 91/414/EEC concerning the placing of plant protection products on the market (94/43/EC). *Official Journal of the European Communities* L 227.

Dushoff, D., Caldwell, B. and Mohler, C.L. (1994) Evaluating the environmental effect of pesticides: a critique of the Environmental Impact Quotient. *American Entomologist*, Autumn 1994.

EUROSTAT (1995) *Europe in Numbers*. EUROSTAT, Brussels, pp. 101, 117, 129.

Faasen, R. (1994) Agricultural pesticide use . . . a threat to the European Environment? Contribution to the Workshop on a Framework for the Sustainable Use of Plant Protection Products in the European Union, 14–15 June 1994, Brussels, Belgium.

Hammond, A., Adriaanse, A., Rodenburg, E., Bryant, D. and Woodward, R. (1995) *Environmental Indicators: A Systematic Approach to Measuring and Reporting on Environmental Policy Performance in the Context of Sustainable Development*. World Resources Institute, Washington.

Harman, H.H. (1960) *Modern Factor Analysis*. University of Chicago Press, Chicago, pp. 301–302.

Heinz, I., Zullei-Seibert, N., Fleischer, G., Schulte-Ebbert, U. and Simbrey, J. (1995) *Economic Efficiency Calculations in Conjunction with the Drinking Water Directive (Directive 80/778/EEC); Part III: The Parameter for Pesticides and Related Products*. Final Report for the European Commission, DGXI, Dortmund.

Hoekstra, J.A. (1996) *Towards an Integrated Environmental Monitoring Strategy*. RIVM Report No. 714701016, Bilthoven.

Koops, R., Linden, A.M.A van der and Berg, R. van den (1996) *Occurrence of Pesticides in Soil. A Pilot Study*. RIVM Report No. 725801009, Bilthoven.

Kovach, J., Petzoldt, C., Degli, J. and Tette, J. (1992) A method to measure the environmental impact of pesticides. *New York's Food and Life Science Bulletin* 139.

Landell Mills (1996) *Regional Analysis of Use Patterns of PPPs [plant protection products] in Six EU Countries: Executive Summary and Cross Regional Reviews: Wheat, Potatoes, Apples, Vines*, Vol. I, (PES-A 2/1). Landell Mills Market Research Limited, Bath.

Mulkey, L.A., Carsel, R.F. and Smith, C.N. (1986) Development, testing, and applications of non-point source models for evaluation of pesticides risks to the environment. In: Georgini, A. and Zingales, F. (eds) *Agricultural Nonpoint Source Pollution: Model Selection and Application*. Elsevier, Amsterdam.

OECD (1994) *The Use of Environmental Indicators for Agricultural Policy Analysis*. COM/AGR/CA/ENV/EPOC(94)48, OECD, Paris.

OECD (1996) *OECD Agri-environmental Indicators: Recent Progress and the Future Programme of Work. Activities of the Joint Working Party of the Committee for Agriculture and the Environment Policy Committee*. Document No 9, OECD, Paris.

OECD (1997) *Environmental Indicators for Agriculture*. OECD, Paris.

Oskam, A.J., Vijftigschild, R.A.N. and Graveland, C. (1998) *Additional EU Policy Instruments for Plant Protection Products*. Wageningen Pers, Wageningen.

Ott, W.R. (1978) *Environmental Indices: Theory and Practice*. Ann Arbor Science Publishers, Ann Arbor.

Pease, W.S., Liebman. J., Landy, D. and Albright, D. (1996) Appendix I. Ranking pesticides by environmental hazard. In: *Pesticide Use in California: Strategies for Reducing Environmental Health Impacts*. University of California/Center for Occupational and Environmental Health, Berkeley, California.

Reus, J.A.W.A. and Pak, G.A. (1993) An environmental yardstick for pesticides. *Mededelingen Faculteit Landbouwkundige en Toegepaste Biologische Wetenschappen Universiteit Gent* 58, 249–255.

SSLRC (1996) *Further Analysis on Presence of Residues and Impact of Plant Protection Products in the EU (PES-A 2/2)*. Soil Survey and Land Research Centre, Cranfield University, Derby.

Theil, H. (1971) *Principles of Econometrics*. Wiley, New York, pp. 47–55.

Van Klaveren, J.D. (1997) *Resultaten Residubewaking in Nederland. Kwaliteitsprogramma Agrarische Producten, Verslag 1996*. RIKILT-DLO, Wageningen, The Netherlands.

Wijnands, F.G and Van Dongen, G.J.M. (1995). Crop protection in Integrated Farming Systems: environmental exposure to pesticides. Paper presented at a Workshop on Pesticides, 24–27 August 1995, Concerted Action AIR3-CT93-1164, Wageningen, The Netherlands.

Wijnands, F.G. and Vereijken, P. (1992) Region-wise development of prototypes of integrated arable farming and outdoor horticulture. *Netherlands Journal of Agricultural Science* 40, 225–238.

Site-specific Water-quality Indicators in Germany

12

Stephan Dabbert, Bernard Kilian and Sabine Sprenger

INTRODUCTION

The problem of water pollution by agriculture is used in this chapter as an example of the development of environmental indicators. Agriculture is a major contributor to water-quality problems in European countries, both with respect to groundwater and surface water. The actual and potential damage to water quality caused by agriculture results from a complex interaction of management practices of farmers and the natural conditions of the site where the farm is located. Practices that are highly dangerous to water quality at one site might have no significant environmental impacts elsewhere.

The driving force pressure–state–response framework (DSR) developed by OECD (Chapter 3) was adapted to show how we conceptualize the problem of water pollution by agriculture (Fig. 12.1). Environmental indicators play a crucial role within this framework, because the actions of different agents depend on them.

The chapter is structured as follows: first there is a discussion of the problems surrounding the choice of adequate environmental indicators; then an indicator approach developed for use in the context of extension and moral persuasion policies is presented, identified by the acronym SINDA (system of site-specific indicators for assessment); in the following section an application of SINDA is demonstrated and the policy implications examined; and finally, some conclusions are drawn.

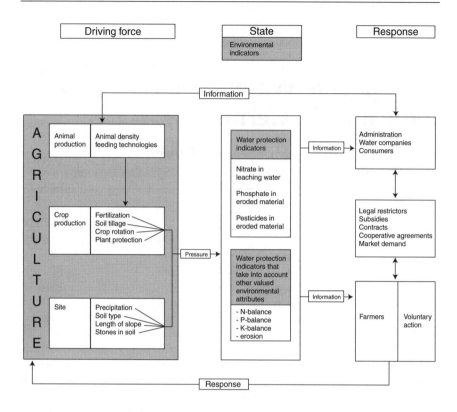

Fig. 12.1. The OECD driving force–state–response framework adapted to the case of water pollution by agriculture.

ENVIRONMENTAL INDICATORS IN AGRICULTURE

Why Do We Need Environmental Indicators?

In the economic analysis of environmental problems, the concept of a physical damage function is useful (Siebert, 1995). For the case of agriculture, the simplest version of such a function would explain damage to the environment (D) by management practices of the farmer (M) and site characteristics (S), with all variables being expressed in physical terms:

$$D = f(M, S).$$

If such a function is known, then the analysis can proceed along the lines outlined in economics textbooks (e.g. Siebert, 1995). However, in many cases it is very difficult to estimate these functions because the underlying causal relationships have been only partly defined by natural scientists. In such a situation, the value of an exogenous variable of the damage function can be taken as an approximation (an indicator) of the total function. This closely parallels the idea that an *indicator* can be used to measure a more abstract *concept*. In our case the concept is the environmental problem (the physical environmental damage). The US Environmental Protection Agency definition of environmental indicators given below is somewhat broader, but does not contradict our earlier statements:

> Environmental Indicator: a parameter (i.e., a measured or observed property), or some value derived from parameters (e.g., *via* an index or model), which provides managerially significant information about patterns or trends (changes) in the state of the environment, in human activities that affect or are affected by the environment, or about relationships among such variables. As defined here, indicators include geographic (spatially referenced) information, and information used in environmental management at any scale, i.e., not just for high-level policy makers.

(US Environmental Protection Agency, 1995)

It is important to note that, because of its nonpoint-source pollution character, the impact of agriculture on water quality cannot be measured precisely on a large scale. Although the amount of nitrate in a specific water source or borehole can be measured, this is not the case with the actual nitrate entering the groundwater associated with particular agricultural practices. Thus, it is important to keep in mind the connotation of an indicator as being something measurable or calculable but, by its very definition, the indicator is not the concept (the *problem*) that is targeted because of its distance from the latter. This implies that indicators are never a perfect quantification of the problem: they are not the full estimation of the damage function but a simplified substitute for it. The most valuable use of indicators in agri-environmental analysis often lies in their ability to facilitate comparison between different situations, rather than as a precise measurement of a problem on an absolute scale.

Examples of Environmental Indicators

The main environmental problem (in terms of physical damage) with which we are concerned here can be defined as 'the contribution of agriculture to the loads of nitrate, phosphate and pesticides in groundwater and/or surface water'. Figure 12.2 shows this problem (concept) and several indicators that have been used to describe the state of the environment with respect to it. Close to the concept are those indicators where the causal chain that lies between them and the concept to be measured is influenced by comparatively few intermediate variables. Far from the concept are those indicators for which the causal chain is longer and where there are more intermediate variables, thus reducing the validity of the indicators. In general, it is the case that the more valid indicators are more difficult and costly to obtain, while readily available indicators are typically less valid. In Fig. 12.2 the indicators are also ordered by a second dimension. The indicators that are strongly influenced by

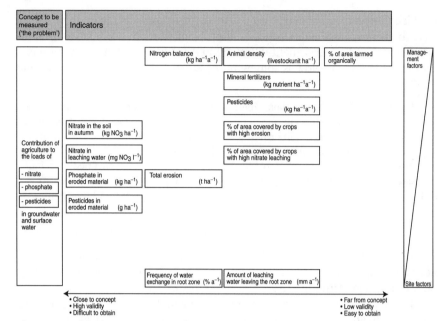

Fig. 12.2. An environmental problem (concept) and different environmental indicators referring to the problem: the case of water pollution by agriculture. a = annum = year.

management factors are listed at the top of the graph, whereas those strongly influenced by site factors are listed towards the bottom. Between these extremes lie those indicators where both site and management are important. This two-dimensional approach obviously introduces a further element of simplification because the attributes *closeness to concept*, *degree of validity* and *ease of measurement* are not always positively correlated.

A second caveat applies: Fig. 12.2 gives examples of indicators that are perceived as important in the German context. However, if the context is expanded, the number of relevant indicators increases. For example, in the case of The Netherlands, phosphate leaching could be added. If pesticides emissions to groundwater occur, then an indicator would have to be added for this.

It is important to note that each of the indicators shown in Fig. 12.2 addresses only part of the problem of water pollution by agriculture. Some indicators are also relevant for other valued environmental attributes. Compare, for example, the indicators *nitrate in leachate* and *nitrogen balance*. The latter also includes nitrogen entering the atmosphere, and therefore extends beyond the scope of the defined problems. Nevertheless, it is less specific in detail, because it indicates nothing about impacts on the valued environmental attributes *groundwater quality* and *atmosphere*. This means that these indicators are actually approximations of different damage functions; that is, they refer to different definitions of what constitutes the environmental problem.

The need for careful selection and interpretation of indicators

The choice of an indicator is not a trivial task but needs to be included into a careful causal analysis of its implications. For example, for the protection of drinking water the nitrate concentration in the leachate leaving the root zone appears to be a relevant indicator of agriculture's contribution to the problem. Under specified management conditions the amount of leachate will determine the nitrate concentration in the leachate. The amount of leachate from the root zone each year could be taken as an indicator for the natural disposition of the site to nitrate leaching: the smaller the quantity of water leaving the root zone, the higher the nitrate concentration. The indicator will show the biggest problems for drinking water on loess soils with low precipitation and the smallest problems on sandy soils with high precipitation.

In some regions, there is very little groundwater recharge below agricultural land, so small amounts of surplus nitrogen will lead to a high concentration of nitrate in the leachate. However, this might not be a serious problem because the absolute load of nitrogen reaching the groundwater is low. It might, in any case, be difficult to use the groundwater for drinking purposes because of the low recharge. Negative effects on surface water will be minimal in this case because the total load reaching the surface water is very low.

The indicator may be further criticized because of its static nature. Nitrate leaching is a dynamic problem. It can be very high for a few weeks of the year, depending mainly on weather conditions (temperature and precipitation), while during the rest of the year very little leaching might occur. This leads some scientists (Deutsche Bodenkundliche Gesellschaft, 1992) to propose the *frequency of water exchange from the root zone* as an indicator for the natural disposition of a site for nitrate leaching from agriculture. The higher this frequency the higher the probability of an actual occurrence of nitrate leaching. The indicator will be highest on sandy soils with high precipitation, and will show relatively small problems on loess soils with low precipitation.

The point here is that the two indicators can lead to apparently opposite conclusions regarding the importance of the natural disposition of a site in contributing to severe nitrate leaching from agriculture if differences in the meaning of the two indicators are not carefully understood. The first indicator looks solely at the nitrate concentration of the leachate, whereas the second is concerned with its potential nitrate load (total quantity). Thus, the first indicator would be useful where there was a requirement that all water leaving the root zone should be of drinking-water quality; the second indicator would be a useful start for further analysis if interest was in limiting the total amount of nitrogen leaving the agricultural system in the direction of the groundwater. Both have been used recently by natural scientists. This points to the need for a very careful interpretation of indicators.

What is a good environmental indicator?

Environmental indicators are part of a more complex problem. They need to be selected in a way that assists in finding an optimal solution to an environmental problem. In theoretical terms the damage function is only a prerequisite for economic analysis. Different levels of analysis are possible. We are interested here in the use of environmental indicators for extension purposes and for policy evaluation. What constitutes a good indicator is consequently dependent upon the purpose of the indicator and the framework of analysis in which the indicator is to be used. If an environmental target is set in physical terms, a good indicator will be one that leads to a least-cost solution to the environmental problem. In different analytical contexts, the types of cost will differ.

Let us first take a look at the extension context for which SINDA (which is explained in more detail later in this chapter) was developed. We start from the assumption that farmers are often unaware that their farming practices contribute to the problem of water pollution. If it can be shown to them in a convincing way that they do contribute, farmers might be more willing to consider alternative practices. In order to be convincing it is generally necessary that the indicators are close to the problem. This approach uses environmental indicators in a moral persuasion context (Nutzinger, 1994). Thus, transaction costs are limited to the cost of extension and do not include any cost for administration and control of policy programmes, nor is there a strong incentive for farmers to behave in such a way that high environmental performance as measured by the indicator is associated with detrimental impacts on other environmental attributes (evasion effects). In this case, the private cost of reaching a given state of the environment can differ if different indicators are used. These costs are, of course, of major interest to the farmer. We hypothesize that these are low if indicators close to the problem are chosen. The cost of obtaining the indicator must also be taken into account when selecting the indicator.

If, however, the indicator is used as a target for policy action (the level of the target being in practice a policy decision) the following types of costs have to be considered (for a somewhat different categorization see Scheele *et al.* 1993):

- the cost of obtaining the indicator;
- the social opportunity cost of production; and
- other transaction costs (including administration and control) associated with the policy implemented.

In this case, the environmental indicator becomes part of a policy instrument, so the question of the best instrument arises (Nutzinger, 1994). The decisions on the indicator and the instrument have to be made jointly in such a way that the target level of the resource is achieved at least cost to society. Politicians do not necessarily subscribe to the preceding view, especially with respect to the use of the social opportunity cost of production. In order to fulfil farmers' expectations they might instead be interested in minimizing the income losses to the farmer. In this case, the private opportunity cost of production might also be of interest when policies are analysed.

A SYSTEM OF SITE SPECIFIC INDICATORS FOR ASSESSMENT (SINDA)

The arguments developed above on environmental indicators were incorporated into the development of SINDA. The primary purpose of SINDA was to provide extension agents with a tool to:

- evaluate the impact of current farming practices on ground- and surface water on individual farms;
- demonstrate the consequences that changes in management practices would have with respect to profit, and the potential impact of the farm on surface and groundwater.

SINDA starts from the hypothesis that farmers lack information about the consequences of improved farming practices in their specific situations. We think that providing a practical way to adapt information to the specifics of a given farm will help to improve the awareness of the farmer both of the problem and of alternative solutions. Of course, if water-protecting agricultural practices seriously decrease profit they will not usually be implemented. However, in a number of case studies we have also found examples where win–win solutions (improved profit, improved environmental protection) exist.

In the following section the rationale for the selection of specific indicators is outlined. The activity level (e.g. production of winter wheat) is the most relevant level for the farm where the aim is to implement changes that will improve water quality. These activities are valued by economic and ecological indicators. The potential danger of arable land use to water quality is described by:

- the nitrogen balance and/or the nitrate concentration in the leachate;
- the quantity of phosphate in eroded material or the quantity of eroded material.

Nitrate Concentration in Leaching Water

Changes in the quantity of nitrogen that is contained in the soil organic matter depend on the type and intensity of land use (especially fertilizer use). Surplus nitrogen resulting from these factors may move downward with the leachate. This potential danger to groundwater depends on the characteristics of the site, and especially on the frequency of exchange of the water in the soil (Becker, 1996). The nitrate concentration in leachate can be used as an indicator for the potential N-load entering the groundwater (Fig. 12.3). From the unsaturated leachate zone the nitrate moves to the saturated zone of the

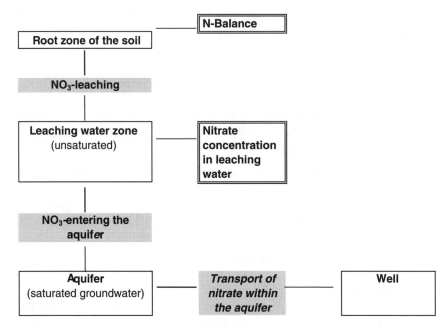

Fig. 12.3. Different zones of nitrate transport in the soil and below (after Rohmann and Sontheimer, 1985).

groundwater (aquifer). Usually, only a proportion of the nitrate entering the groundwater will finally reach a water source or borehole. The magnitude of this proportion depends on the distance and the specific hydrogeological circumstances (Rohmann and Sontheimer, 1985).

The nitrate concentration within an aquifer is thus the consequence of numerous factors. If a high concentration is observed, this means that only general conclusions can be drawn and it is impossible to pinpoint specific farming activities as being directly and solely responsible. However, if the contribution of farming to the problem is to be assessed, specific actual or possible farming activities have to be evaluated. Such an evaluation is possible with the help of the environmental indicator *nitrate concentration in leaching water*. An alternative would be to use the N-balance (we use field balances throughout) as an indicator, which is somewhat more distant from the problem but still convincing to the farmer.

Phosphate in Eroded Material versus Quantity of Eroded Material

In order to reduce the phosphate load in surface water, an indicator is needed as a managerial tool that reflects the impact of agricultural land use on surface water. The quantity of phosphate in our model depends on total erosion and a factor that represents the selective transport of different soil particles (Auerswald, 1989).

Total soil erosion is of special importance here because the phosphate content of the soil changes only in the medium to long run. Even with surpluses of phosphate there are not always phosphate emissions. For this reason, phosphate balance is not a valid indicator for the problem. Soil erosion depends on site factors and management (e.g. tillage and crop rotation). Soil erosion is calculated with the help of the universal soil loss equation (USLE) (Wischmeier and Smith, 1978) or, more precisely, the version of the USLE that has been adapted to German circumstances – Allgemeine Bodenabtragsgleichung (ABAG) (Schwertmann *et al.*, 1990) – is used to provide an indicator of potential damage to surface water.

The indicator quantity of phosphate in eroded material can now be used to evaluate different land-use activities with respect to their (potential) influence on surface-water quality (Fig. 12.4). If the phosphate load of the surface water only were measured, such conclusions would be impossible.

It is somewhat simpler to calculate the total quantity of eroded material. Less information is needed, so this indicator is cheaper to obtain. For a given site this indicator is positively correlated with the indicator *phosphate in eroded material* and is thus a better choice in many cases.

SINDA provides an extension agent with examples of different farming activities (e.g. winter wheat). These examples are the result of intensive discussions with extension agents and then applications to specific sites in specific regions. The examples are differentiated by the factors most important for the impact of agriculture on ground- and surface water: yield level, fertilization and soil tillage. These examples can then be adapted to the circumstances of a specific farm. The details on how to calculate the indicators are not elaborated here but can be found elsewhere (Frede and Dabbert, 1998).

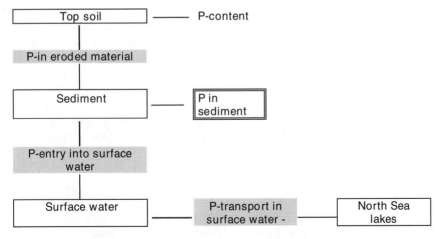

Fig. 12.4. Different zones of the transport of phosphate into the surface water.

APPLICATION OF SINDA

A farm-level economic–ecological model was developed that incorporated the SINDA concept. Different sites and farm types were selected carefully in order to include the range of natural sites (sensitivity to nitrate leaching and erosion ranging from very low to very high) and farm types found in Germany. The model is a very disaggregated linear programming model based as much as possible on empirical data from the specific case studies. The conventional profit maximization assumption applies.

A schematic and simplified overview on the structure of the matrix used is given in Fig. 12.5. This shows that important variables are calculated endogenously within the model. An example is the c-factor of USLE which in most linear models is a parameter calculated beforehand, while in our model it is a variable. In the following section the model is used to analyse the economic effects of a more crude environmental indicator in comparison to a more precise environmental indicator for the case of nitrate in groundwater, using an intensive livestock farm at a site with a high potential for nitrate leaching as an example.

Farm gross margin, farm organization and the state of the environment (as measured by environmental indicators) are calculated for three cases. In the first case, no environmental restrictions are applied. In the second, the nitrogen surplus (as measured by the nitrogen balance) is restricted to zero, thus triggering changes to farm organization and fertilizer use. In the third case, a different environmental restriction is imposed on the model and the indicator used is nitrate concentration in leachate in the root zone: a nitrate concentration close to the natural basic load is allowed and the results of the optimizations are shown in Table 12.1.

In the first case, with no environmental restriction, the nitrate concentration in the leachate indicates an environmental problem. Cases 2 and 3 lead more or less to the same results with respect to the state of the environment. In both cases no negative impact of agriculture on groundwater quality would be expected. Also it is clear that, from the farmer's view (assuming perfect information and no monitoring cost), the indicator that is closer to the problem is much to be preferred for economic reasons. It is much cheaper to reach the desired environmental state in terms of opportunity cost if the indicator closer to the problem (nitrate concentration) is targeted. While this is a deduction from one example, we suggest that it will hold in most cases: the closer the indicator is to the problem, the lower are the private opportunity costs of foregone production if a comparable state of the environment is to be achieved.

A second example serves to apply SINDA to policy evaluation. Here the example of an arable farm at a site with high erosion danger is taken (Table 12.2). In cases 4 and 5 the agricultural policy framework is applied as it currently exists. The cases differ with respect to the tillage system used on the farm; in all other aspects profit-maximizing behaviour is assumed. The results

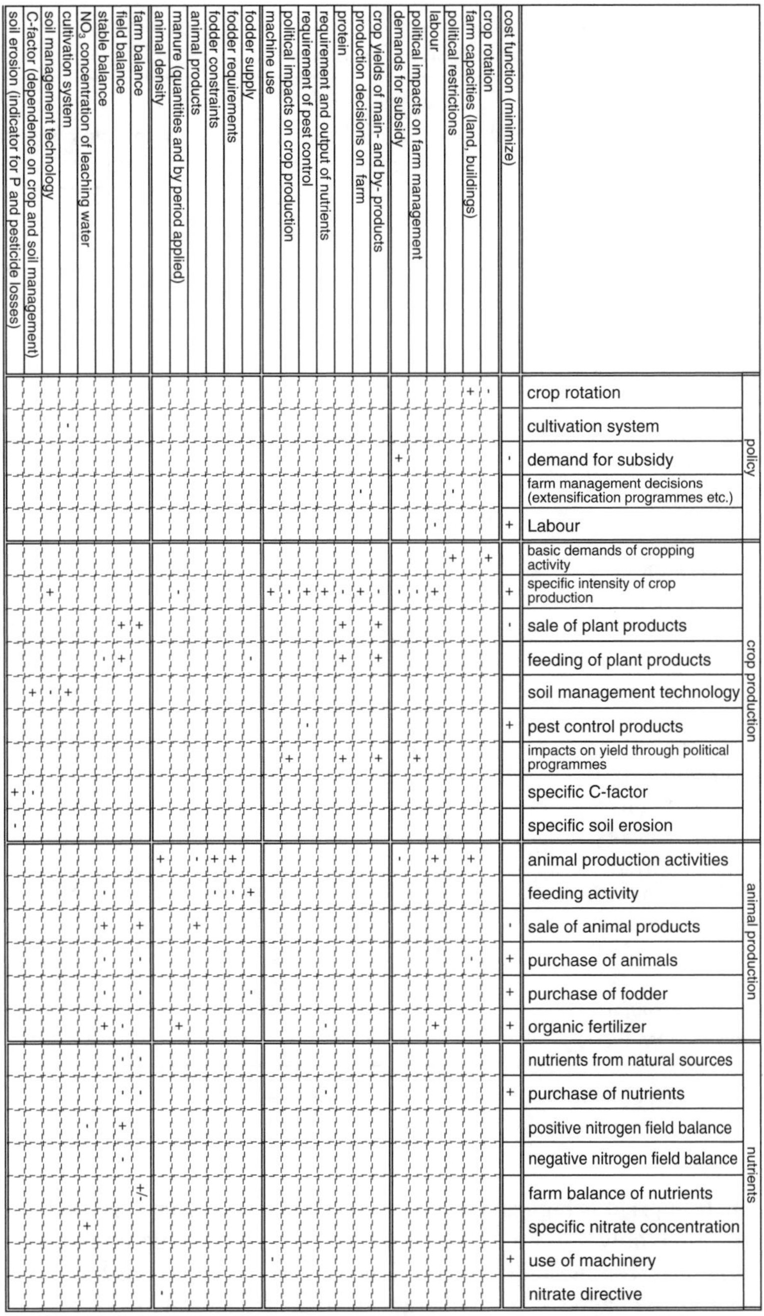

Fig. 12.5. Schematic overview on model structure; + and − indicate links between activity and restrictions: −, activity supplies into restriction; +, activity demands from restriction.

Table 12.1. The impact of different environmental restrictions aiming at groundwater protection on an intensive livestock farm.

	Case 1 without environmental restrictions		Case 2 N balance = 0		Case 3 nitrate concentration < 8 mg l⁻¹	
	ha	kg N ha⁻¹	ha	kg N ha⁻¹	ha	kg N ha⁻¹
Land use						
Arable land	60		60		60	
Crops for industrial use	6	0	6	0	6	0
Wheat	30	210	30	150	30	185
Winter barley	12	150			12.43	120
Spring barley	0.6	105				
Triticale	0.6	70	7	70	0.42	70
Oilseed rape			4.9	5		
Maize	10.8	170	12.1	55	11.15	130
Animal production						
Fattening pig (places)	720		662		720	
Livestock density (LU ha⁻¹)	1.44		1.32		1.44	
Environmental indicators for water protection						
C-factor (of ABAG)	0.15		0.14		0.14	
Soil erosion (t ha⁻¹)	1.5		1.4		1.4	
N-balance (kg ha⁻¹)	57				37	
Nitrate in leaching water (mg l⁻¹)	52				8	
N-fertilization (kg/ha)	170		95		142	
Economic success						
Gross margin (DM/farm)	152 900		137 870		151 610	
	100%		90%		99%	
Subsidies (DM/farm)	36 121		39 010		36 208	
Subsidies (DM ha⁻¹)	602		650		603	

ABAG = Allgemeine Bodenabtragsgleichung; LU, livestock units.

indicate that profit is highest if conservation tillage (mulching) is used. In this case erosion is only 9.6t ha⁻¹ compared to 33.6 t ha⁻¹ if ploughing is used.

In case 6, the agricultural policy framework is kept constant with the exception of the Baden-Württemberg version of EU regulation 2078/92 (CEC, 1992), known as MEKA (Marktentlastungs- und Kulturlandschaftransgleich). This programme has been treated as being non-existent in case 6. The model is free to choose the most profitable tillage system. Profit is then very similar to that under the standard ploughing system, as is erosion. The reason for these

Table 12.2. The impact of different tillage systems and a change in the policy framework on erosion: the case of an arable farm.

	Case 4 ploughing		Case 5 mulch seeding		Case 6 without 2078/92	
	ha	kg N ha^{-1}	ha	kg N ha^{-1}	ha	kg N ha^{-1}
Land use						
Arable land	60		60		60	
Crops for industrial use	4.2	0	4.2	0	4.2	0
Wheat	30	225	30	225	30	225
Spring barley			0.8	105		
Maize	7.8	170	7	170	7.8	170
Sugarbeet	18	145	18	145	18	145
Catch crop	26		26			
Environmental indicators for water protection						
C-factor (ABAG)	0.28		0.08		0.26	
Soil erosion (t ha^{-1})	33.6		9.6		31.2	
N-balance (kg ha^{-1})	32		32		28	
Nitrate in leaching water (mg l^{-1})	24		24		21	
Economic success						
Gross margin (DM farm^{-1})	141 022		149 858		141 210	
Subsidies (DM farm^{-1})	31 118		35 982		25 345	
Subsidies (DM ha^{-1})	519		573		422	
MEKA (DM farm^{-1})	5 772		10 812			

ABAG, Allgemeine Bodenabtragsgleichung; MEKA, Marktentlastungs- und Kulturlandschafttsaugleich.

results is a provision in MEKA for paying mulching subsidies. It can thus be concluded for the study farm that these provisions actually lead to an improved profitability of the mulching system. MEKA thus has a strong positive environmental impact. This result is supported by empirical evidence (Zeddies and Doluschitz, 1996) that shows a considerable increase of the use of the mulching system with the introduction of the MEKA in Baden-Württemberg. However, the modelling approach makes the causal connection more evident.

The example presented makes it clear that the combination of a linear farm model with carefully selected environmental indicators can be a powerful tool for evaluating the impacts of agricultural policies on the environment. A more complete evaluation would clearly require a larger number of farm cases to be investigated.

CONCLUSIONS

Environmental indicators can provide valuable information to farmers and policy-makers. Some of the secondary indicators in use (animal density, amount of fertilizer used, percentage of area farmed organically, etc.) are rather crude. Even so, dividing the environmental impact of farming into many sub-categories and searching for the indicators that most closely match these sub-problems might not be a feasible strategy. The cost of data collection could be prohibitive. The negative side-effects of an improvement of an indicator of one sub-problem on other sub-problems may be a second disadvantage of this approach.

The central argument here is that environmental indicators have to be selected very carefully and according to least-cost criteria. The types of cost to be considered differ according to the analytical framework used and the level of analysis. For extension purposes our evidence supports the proposition that the private opportunity cost of foregone production will be lowest if indicators close to the problem are chosen. However, the cost of obtaining such indicators is likely to be high, although the total cost (private opportunity cost of foregone production and cost of obtaining the indicator) would be typically less than would be the case if a cruder indicator were to be used. Consequently we suggest (and present) for extension purposes the use of indicators more towards the left of Fig. 12.2.

Costs for the provision of information as well as the costs associated with monitoring and administration (Whitby *et al.*, 1998) also need to be considered in a policy context if environmental indicators are used for specifying farm-level targets within policy instruments. Also the cost of evasion effects must be taken into account as other valued environmental attributes might deteriorate. While quantitative information on these costs is difficult to obtain, we believe that they are low if a small number of carefully selected but crude indicators are used as policy targets. This implies that good indicators for targeting policies might be found more in the middle or towards the right of Fig. 12.2. This hypothesis should be subject to further empirical investigation.

This chapter has concentrated on water quality. Though not without its own complexity, it may be more difficult to find valid indicators for other valued environmental attributes, such as biodiversity. A shortcoming of this approach is that it is not easy to convert it to a system that covers all areas in a country or the EU; for such an approach more refined methods that make use of available secondary data are needed. Research is under way to provide the methodological basis for this.

ACKNOWLEDGEMENT

The research reported in this chapter has been funded by the Umweltbundesamt. The chapter is based on work that was carried out in close cooperation with the University of Gießen (Prof. Dr H.-G. Frede, Dr N. Feldwisch) and that has been documented in Frede and Dabbert (1998). The authors of the paper are solely responsible for its content.

REFERENCES

Auerswald, K. (1989) Predicting nutrient enrichment from long-term average soil loss. *Soil Technology* 2, 271–277.

Becker, R. (1996) Regional differenzierte Bewertung von Maßnahmen zur Minderung von Stickstoffüberschüssen mittels Stickstoff Bilanzen. Dissertation. Justus-Liebig Universität Gießen.

CEC (1992) Council Regulation (EEC) No 2078/92 on agricultural production methods compatible with the requirements of the protection of the environment and the maintenance of the countryside. *Official Journal of the European Communities* L 215 (30/7/92), 85–90.

Deutsche Bodenkundliche Gesellschaft (1992) *Strategien zur Reduzierung standort- und nutzungsbedingter Belastungen des Grundwassers mit Nitrat*. Arbeitsgruppe Bodennutzung in Wasserschutz- und schongebieten, Oldenburg.

Frede, H.-G. and Dabbert, S. (eds) (1998) *Handbuch zum Gewässerschutz in der Landwirtschaft*. Ecomed, Landsberg, 450 pp.

Nutzinger, H.G. (1994) Economic instruments for environmental protection in agriculture: some basic problems of implementation. In: Opschoor, H. and Turner, K. (eds) *Economic Incentives and Environmental Policies*. Kluwer Academic Publishers, Dordrecht, pp. 175–193.

Rohmann, U. and Sontheimer, H. (1985) *Nitrat im Grundwasser – Ursachen, Bedeutung, Lösungswege*. DVGW-Forschungsstelle am Engler-Bunte-Institut der Universität Karlsruhe, Frankfurt am Main, Germany.

Scheele, M., Isermeyer F. and Schmitt, G. (1993) Umweltpolitische Strategien zur Lösung der Stickstoffproblematik in der Landwirtschaft. *Agrarwirtschaft* 42, 294–313.

Schwertmann, U., Vogl, W. and Kainz, M. (1990) *Bodenerosion durch Wasser: Vorhersage des Abtrags und Bewertung von Gegenmaßnahmen*, 2nd edn. Ulmer, Stuttgart.

Siebert, H. (1995) *Economics of the environment. Theory and Policy*. Springer, Berlin.

United States Environmental Protection Agency (1995) *Conceptual Framework to Support Development and Use of Environmental Information in Decision Making*. Document Number 239-R-95-012. US Environmental Protection Agency, Washington, DC.

Whitby, M., Saunders, C. and Ray, C. (1998) The full cost of stewardship policies. In: Dabbert, S., Dubgaard, A., Slangen, L. and Whitby, M. (eds) *The Economics of Landscape and Wildlife Conservation*. CAB International, Wallingford, pp. 97–112.

Wischmeier,W.H. and Smith, D.D. (1978) *Predicting Rainfall Erosion Losses - A Guide to Conservation Planning*. USDA, Agricultural Handbook No. 537. USDA, Washington DC.

Zeddies, J. and Doluschitz, R. (1996) *Marktentlastungs- und Kulturlandschaftsausgleich (MEKA); Wissenschaftliche Begleituntersuchung zu Durchführung und Auswirkungen*. Ulmer, Stuttgart.

Nutrient Balances and the Implementation of Agricultural Policy Measures in Finland

13

Reijo Pirttijärvi, Seppo Rekolainen and Juha Grönroos

INTRODUCTION

Background

The extent of water-quality problems caused by nonpoint sources of pollutants, and particularly those emanating from agricultural activities, highlights the need for policy intervention. Nutrient balance calculations (NBCs) provide a method for assessing the nutrient surpluses of specific farms, and hence their environmental loads. Using this approach it is also possible to assess the effect of agri-environmental policy measures on nutrient runoff. However, NBCs are not without their problems: these will be scrutinized in this chapter.

The accession of Finland to the European Union (EU) in 1995 modified Finnish agriculture. The subsidy system of the pre-EU era had to be replaced by one based largely on direct payments instead of price support, and one which incorporated a significant environmental component. Measures introduced under the EU Agri-Environmental Policy are regarded as a very important way of promoting environmental protection, through the prevention of environmental degradation and the production of environmental benefits. Nutrient balances provide one way in which the impact of the programme on the environment can be assessed.

The Objective of the Study

The main objective of this chapter is to present nitrogen balance information relating to Finland, to discuss the problems in using it for assessing

©CAB INTERNATIONAL 1999. *Environmental Indicators and Agricultural Policy* (eds F.M. Brouwer and J.R. Crabtree)

environmental pressures from agriculture, and to give an overview of the expected environmental benefits of agri-environmental support in Finland.

NUTRIENT BALANCES

Nutrient Balance Calculations

NBCs (which are also referred to as mineral balances and nutrient budget calculations) have been applied extensively as environmental indicators during the past decade. The main use of NBCs has been to estimate nutrient surpluses at farm, regional or national level. Estimation methods have varied, reflecting, in part, a range of different objectives.

The nutrient flows of an agricultural system are usually expressed either by the farm-gate balance or by the surface balance. The idea behind the *farm-gate balance* is to measure how much purchased nutrient inputs are used on a farm in its production, and to compare these figures with the nutrient contents of the outputs leaving the farm. The level of observation is at the actual inflows and outflows of nutrients crossing the farm's boundary ('the farm gate'). For example, in the context of nitrogen and phosphorus, the quantities of the inputs entering the farm are compared with the corresponding figures exiting the farm in the outputs sold, the difference giving a surplus or a deficit of nutrients.

The *surface balance* is calculated as the difference between the nutrients entering into and exiting from the soil surface. The input calculation takes into account the use of chemical fertilizers and manure, whereas that for outputs incorporates the nutrients in harvested crops. Subtracting the latter from the former gives a gross surface balance. The net surface balance also takes into account the evaporation of nitrogen from manure and the fact that not all the nitrogen in manure can be utilized by plants. Usually a 30% estimate for non-available nitrogen is subtracted from the nitrogen in the manure (e.g. Brouwer *et al.*, 1995). For phosphorus the gross balance is satisfactory without further adjustment since phosphorus losses from evaporation are negligible. Both calculation methods can incorporate atmospheric deposition and biological nitrogen fixation, the latter being particularly important with organic farming systems.

Each method yields a farm-specific nutrient surplus or deficit. Efficiency of nutrient use can be calculated by comparing the nutrient output to its input. While, in theory, both calculation methods should yield the same result, this is not usually the case in practice. The difference probably stems from the use of standard coefficients when assessing the nutrient contents of animal manure, crop production and livestock production (Brouwer *et al.*, 1995). The nutrient surplus indicates the short-run pollution potential of emissions. In the long run, the mean nutrient surplus is an approximate measure of the farm's environmental load.

Experience of Nutrient Balances

The Organisation for Economic Co-operation and Development (OECD) has developed national nitrogen balance estimates using the surface balance method which takes into account ammonia losses from manure before spreading. Figure 13.1 shows the Finnish nitrogen surplus according to the OECD method over an 11-year period. As can be seen from the figure, there is a downward trend in the nitrogen surplus during this period. The slope of the regression curve estimated from the series and shown in the figure is −3.0, i.e. nitrogen surplus per hectare is decreasing, on average, by approximately $3 \ kg \ ha^{-1} \ year^{-1}$, although in the last few years there has been an upward trend. The same kind of a general trend is taking place in most other OECD countries (OECD, 1997).

Another feature evident from Fig. 13.1 is that the surplus in 1987 deviated considerably from the trend line. This was due to poor weather conditions for crop growth during that year. Nitrogen applications via fertilizer and manure were about the same as in 1986, but the uptake of nitrogen by the crops was about 30% lower. The consequence in 1987 was a nitrogen surplus almost 20 $kg \ ha^{-1}$ higher than in 1986.

This kind of weather-induced deviation in nutrient losses has important implications at policy level. Had a policy measure been established in 1986 with the objective of reducing the nitrogen surplus, the 1987 balance figure could have been incorrectly interpreted as an indication of policy failure. This is important in an international context where uniform policies are often desired, and national characteristics often not well understood.

This type of problem could be avoided by calculating, for example, a 3-year moving average, but then the effect of a particular policy becomes difficult to pinpoint due to the smoothing. Alternatively, some sort of 'tolerance limits' for the random factor in the nutrient surplus could be introduced. However, there is no unambiguous way to set tolerance limits as the annual changes can be considerable. If set too wide, the effect of a policy measure will again be masked.

Some 500 pork-producing farms are involved in production planning organized by the Centre of Rural Advisory Services (MKL). In the MKL production planning system nutrient balances were calculated in 1995 (Helander, 1996). The relationship between production costs and the percentage utilization of nutrients is illustrated in Fig. 13.2.

The nutrient balance calculated with MKL data depicts the relationship between the nutrients entering the herd and exiting it in the form of meat. (This is termed the feed nutrient balance.) There is a positive correlation between nutrient utilization and production costs; in other words, the lower the production cost the better is the farm's utilization of nutrients. This demonstrates a clear link between good farm management and good environmental management.

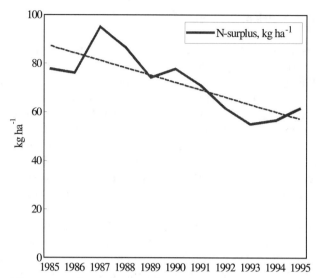

Fig. 13.1.   Nitrogen surplus and its trend in Finland in 1985–1995.

THE USE OF NUTRIENT BALANCE CALCULATIONS

Inherent Problems of Nutrient Balance Calculations

Even though the NBCs seem to be a very attractive way of circumventing the nonpoint pollution problem, thus allowing the policy-maker to assess the situation from a point-source perspective (and to use economic instruments such as levies on nutrient surpluses), some inherent limitations still exist. These limitations can be calculation specific, information specific or application specific.

The *calculation-specific problems* are those that relate to data accuracy and to the elements that need to be included in the calculation. In addition, the level (regional, national or international) at which the calculation is made is important because this may determine the coefficients for the nutrient contents in the manure and harvested crops used in the surface balance calculation. Estimation of nutrient coefficients for the nutrient content in the manure of different animals is always problematic. The contents can change considerably between years and differ regionally and internationally. The magnitude of the coefficients depends primarily on feeding regimes and digestive efficiency. The same kind of variation, but to a lesser extent, is present in the nutrient uptake coefficients of different crops.

Although biological nitrogen fixation is the cornerstone of the nutrient economy in organic farming, in most cases it makes a negligible contribution to conventional farming. Nitrogen deposition is also subject to wide variation. It can be as high as 35–40 kg ha^{-1} in central Europe, but in Portugal or in

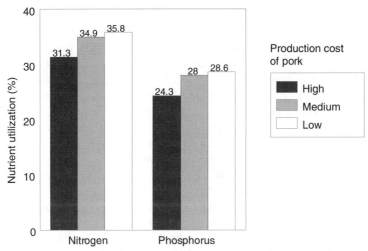

Fig. 13.2. The relationship between production costs and nutrient utilization on Finnish pork farms in 1995. Source: Helander (1996).

Finland is typically around 5 kg ha^{-1} (Brouwer *et al.*, 1995). This divergence is relevant when making international comparisons. Thus, the type of farming and also local conditions must be taken into account when assessing the relevance of factors to be included in the NBCs.

One of the major questions concerning the feasibility of using mineral balances for policy determination is the problem of external factors, such as weather and soil type, which affect the accommodation of a nutrient surplus. Yield levels vary from year to year due to weather effects and this is particularly marked in Finland which is at the margin of cultivation for many crops. The effect of climate on nutrient balances is considered in greater detail below.

In some calculations (Schleef and Kleinhanß, 1994; Brouwer *et al.*, 1995) 30% of the nitrogen in manure is deducted in the calculation of net surface balance, as explained above. Calculating a net surface balance for nitrogen tends to reduce any differences between cereal and livestock farms in the efficiency of nutrient use. The net balance is also more closely related to the problem of nitrogen losses to watercourses. However, this approach distorts the real problem of agricultural nutrient losses, as losses to the air and soil are ignored. Also, at the farm level the 'evaporation percentage' may well differ from 30% due to the differences in production techniques.

There are also *information-specific problems* involved in the NBCs. In general, nutrient balances are not very useful indicators unless related to agricultural land area or the total input of nutrients. In the international context the choice of the elements in the denominator plays an important role. In adding all possible agricultural land to the denominator, the balance per hectare gets smaller (i.e. the environmental impact is lower). However, restriction of the agricultural land area to the active field area (where manure or fertilizers are spread) produces more meaningful indicators.

Furthermore, the NBCs do not indicate precisely where the nutrient emission occurs, i.e. to what extent the losses leach to watercourses or discharge to air. The nutrient balance is thus more of an indicator than an exact measure of nutrient emission. Substantial nutrient surpluses indicate to a farmer that there is both a production efficiency problem in using the inputs at his disposal and a resultant environmental problem. The farmer then faces a number of options for reducing nutrient losses. From this perspective, the NBCs can be seen as an information tool in production planning.

NBCs alone are usually insufficient for comprehensive decision-making. Although they will encourage farmers to select environmentally sound management practices, the balance sheet needs to be integrated closely to other farm planning and monitoring procedures, such as soil analysis.

The question of the purpose and of the actual use of the NBCs relates to the set of *application-specific problems.* The NBCs can be used for different purposes, and it is the purpose which should dictate the calculation method for the balance. If the NBCs are used only as a production planning tool in farm management, a comprehensive approach based on the use of best estimates is more appropriate than strict administrative comparability (Lord and Anthony, 1997). However, if comparability of NBC is the aim, the calculation method must be defined such that the resulting database is consistent.

When used for regulation purposes the balance calculations face a problem of reliability bias due to the strategic behaviour of farmers. If farmers know that the NBCs are to be used, for example, to set a levy based on the nutrient surplus, they will tend to overestimate the quantity of outputs and underestimate the quantity of inputs. The administrative costs associated with verifying the NBC information, with information from a taxation database, for example, will not be insignificant.

Hence, the actual use of NBCs will influence the accuracy of the data. If farmers consider that NBCs provide them with new and valuable information, they have an incentive to calculate the balance sheet as accurately as possible. But, if they risk a penalty depending on the outcome of the calculation, they will have an incentive to report less accurately.

The Importance of Climatic Factors

The importance of external factors in the formation of the mineral surpluses is often raised when assessing the use of the NBCs (e.g. Lankoski, 1996). If external factors largely determine the level of the nutrient balance, there is less justification for setting a levy based on the surplus. Mukula and Rantanen (1989) have studied the variability in the yield of field crops in Finland. The estimated coefficients of variation (the standard deviation expressed as a percentage of the mean yield) were 15% for barley, 19% for winter wheat, 16% for spring wheat, 11% for barley and 9% for oats.

The yield levels are closely linked to nutrient balances, especially with nitrogen balance where the correlation exceeds 0.9. Hence, the coefficients of variation also indicate the magnitude of the climate-induced variation in nutrient balance. An analysis of more recent Finnish time-series data is shown in Table 13.1. The nitrogen surplus has a very high standard deviation, especially with potatoes, which indicates a high variance in the actual crop uptake of nitrogen. As regards the mean level of nutrient surpluses, rye produces a very high nitrogen surplus (almost 91 kg ha^{-1}) whereas potatoes are associated with a high level of phosphorus emission (88 kg ha^{-1}).

The between-year variability in nutrient surpluses is high and especially so for nitrogen. Even though potatoes have a high efficiency of nitrogen utilization, the crop is very susceptible to adverse weather conditions, and summer night frost may result in almost total loss of the crop. This occurred in a few cases and the low yields (less than 15 t ha^{-1}) contributed to very high nutrient losses. This is also true for oats, where night frost can cause a dramatic decrease in yield.

In further analyses, correlation matrices for each of the four crops were first calculated to see whether there was a linear correlation between nutrient surpluses on the one hand, and precipitation and effective temperature sum (ETS) on the other. Then, variance analyses were performed to identify the relevant weather factors associated with the formation of nutrient surpluses. The statistically significant correlations at $P < 0.05$ are shown in Table 13.2. The level of rainfall in May is an important factor in the build-up of the nutrient losses, with excess rain in May contributing to a high nutrient surplus. In other months the precipitation correlations are mostly positive, i.e. the higher the rainfall the larger the nutrient surplus.

The temperature in August is also important for the build-up of nutrient surpluses, as the very consistent negative correlations indicate. The higher the degree-days in August the lower is the nutrient surplus and the more efficient the utilization of nutrients. The same holds true to some extent for the ETS in May.

With few correlations higher than 0.4, we can conclude either that climatic factors are only weak determinants of nutrient surplus or that the relationship between, for example, precipitation and nutrient surplus is not linear. Unusually low or high rainfall in, for example, May probably has similar effects on the nutrient balance, that is, extreme rainfall conditions contribute to high nutrient surpluses. This relationship is therefore almost certainly non-linear, as noted by Teigen and Thomas (1995) who postulated that the climate is the single most important factor affecting crop production. They presented models showing weather–yield response which explains more than 90% of yield variation.

Table 13.1. Mean, standard deviation and variation coefficient for nutrient surpluses by research station and crop.

Research station	Crop	Number of observations	Nitrogen			Phosphorus		
			Mean	Standard deviation	Coefficient of variation (%)	Mean	Standard deviation	Coefficient of variation (%)
Satakunta	Rye	18	84.7	12.3	15	30.2	7.5	25
	Barley	36	35.0	21.5	61	27.3	3.7	14
	Oats	36	18.4	16.5	90	24.0	4.5	19
	Potatoes	18	28.5	18.8	66	89.7	7.6	8
Sata-Häme	Rye	18	93.8	14.4	15	31.2	6.2	20
	Barley	33	54.2	15.7	29	30.6	4.7	15
	Oats	30	38.1	15.3	40	28.3	4.9	17
	Potatoes	17	11.2	30.9	276	87.7	9.6	11
Etelä-Pohjanmaa	Rye	18	84.4	11.2	13	29.5	5.5	19
	Barley	36	38.7	12.4	32	26.8	4.7	18
	Oats	36	9.7	16.7	172	22.4	5.4	24
	Potatoes	18	19.8	26.1	132	88.6	7.0	8
Karjala	Rye	18	99.9	18.5	19	32.3	4.7	15
	Barley	33	43.0	20.4	47	28.1	5.4	19
	Oats	30	27.3	26.3	96	26.0	6.8	26
	Potatoes	18	15.8	30.3	192	86.2	8.6	10
Means	Rye	72	90.7	15.5	17	30.8	6.0	19
	Barley	138	42.5	19.1	45	28.1	4.8	17
	Oats	132	22.5	21.5	96	25.0	5.8	23
	Potatoes	71	18.9	27.1	143	88.0	8.2	9

Table 13.2. Statistically significant correlations (at $P < 0.05$) between nutrient surplus and weather factors in May–August.

Crop	Nutrient surplus	Precipitation (May–August)					ETS (May–August)				
		Month					Month				
		5	6	7	8	5–8	5	6	7	8	5–8
Barley	N						−0.28				
	P	0.36	0.18		−0.17				0.17	−0.20	
Oats	N	0.34		0.27	0.40	0.40		0.32	0.24	−0.29	0.17
	P	0.52		0.21		0.27		0.46	0.30	−0.34	0.28
Rye	N		0.38	0.33	0.61	0.68		−0.30		−0.37	−0.33
	P			0.35	0.23	0.29	0.34			−0.39	0.37
Potatoes	N	0.50	0.35				−0.40			−0.34	
	P	0.58	0.38				−0.31			−0.26	

THE FINNISH AGRI-ENVIRONMENTAL PROGRAMME

Overview of the Programme

The Finnish Agri-Environmental Programme was prepared by the Ministry of Agriculture and Forestry and the Ministry of the Environment, and was adopted by the Commission in late 1995. The objectives and the measures of the programme are based on Regulation 2078/92 (EEC, 1992), on the national governmental decisions for agricultural environment protection and on the international agreements in connection of the Helsinki Commission (HELCOM) Convention with the Protection of the Marine Environment of the Baltic Sea (1974 and 1992). The aim of the Agri-Environmental Programme is to maintain or to direct agriculture in Finland towards a more sustainable basis. The programme is designed to promote environmentally sound agricultural practices.

The main objectives of the programme are to reduce erosion and nutrient losses to water, to reduce atmospheric emissions, to protect and enhance the biodiversity of agricultural landscapes and to mitigate the harmful effects of pesticides. The programme consists of three schemes: the General Agricultural Environment Protection Scheme (GAEPS), Supplementary Protection Scheme (SPS), and educational and demonstration projects. The GAEPS covers the whole of Finland and an annual premium per hectare is paid to farmers who comply with a set of conditions. The premium is differentiated by region. The SPS includes additional measures which apply only in designated areas or in certain horizontal programmes such as organic production.

In order to be eligible for the GAEPS the farmer has to conform to the following conditions:

1. a Farm Environmental Management Plan must be prepared;
2. the maximum levels of fertilization must not be exceeded;
3. manure must be appropriately stored and may not be spread on frozen soil or on snow;
4. livestock density may not exceed 1.5 livestock units (LU) ha^{-1};
5. buffer strips (width 1–3 m) must be established on the sides of main ditches and water courses;
6. at least 30% of arable land must be covered by crops or crop residues during the winter in southern Finland; and
7. pesticide spraying devices must be tested by an authorized agency and pesticides may be applied only by a person who has completed training on pesticide use.

When applying for the SPS the farmer must also participate in the GAEPS. The most important SPS measures are:

- organic production;
- establishment of wide (*c.* 15 m) riparian zones along rivers;
- treatment of runoff waters from arable land, e.g. by establishing sedimentation ponds or wetlands;
- landscape management and biodiversity enhancement;
- maintenance of local breeds in danger of extinction.

Riparian zones and treatment of runoff waters are restricted to focal areas of water protection and groundwater formation defined by environmental authorities. The landscape management and development measures are limited mainly to valued landscape areas. The most important SPS measure in monetary terms is the support to organic farming. In 1995, 70 million Finnish marks (FIM) were allocated to this measure. Prior to the programme, about 30 000 ha were cultivated organically. It is estimated that by the end of 1999 over 70 000 ha will be under organic production, and additionally 45 000 ha will be under conversion to organic production.

Environmental Impacts of the Programme

A priori impact assessments
Environmental impacts of the programme were assessed a priori when preparing its objectives. It was expected that the programme would result in more extensive agriculture. As a consequence there would be a decrease in the environmental load, and, thus, in improved water and air quality, as well as increased biodiversity and beneficial impacts on the quality of agricultural landscapes.

More specifically, the expected changes in the quality of the environment, as identified by changes in agri-environmental indicators, were:

• the use of fertilizers and pesticides would decrease;
• the use of nutrients in manure would become more efficient;
• organic production was expected to increase;
• the employment of reduced tillage techniques would increase;
• buffer strips would be established along the waterways and main ditches;
• the deterioration of agricultural landscapes would be prevented; and
• the loss of genetic material and biodiversity in agricultural and agriculture-related ecosystems would cease.

It was estimated that, in the long run, the adoption of the Agri-Environmental Support Scheme would decrease both erosion and phosphorus and nitrogen losses into watercourses by about 25–40%. In addition, the improved management of animal manure would reduce the contamination problems in wells and small catchments, and decrease ammonia emissions from agriculture. No single measure alone is effective enough to achieve the targeted 50% reduction in diffuse nutrient losses from agriculture set by international organizations (e.g. HELCOM). The anticipated 25–40% reduction will be achieved only if a large majority of farmers join the GAEPS, and if the SPS measures can be focused in designated areas, where their effectiveness is highest. It should be noted that the desired results can be expected only after a relatively long time because the natural processes involved in both soil and water are slow.

Participation in the schemes
Experience from the first year of the programme showed that most of the farmers have undertaken the GAEPS preconditions. However, fulfilment of the conditions has not been satisfactory in all aspects. Moreover, the SPS programme was not implemented as originally planned. This means that, if no clear improvement takes place in the coming years in its implementation, the original goals will probably not be achieved.

The GAEPS attracted more participants than originally expected. Some 80 000 farms in total joined, leaving only about 20% of all farmers not participating. The average rate of participation was highest (89% of all farms) in southern Finland where the premiums are higher than elsewhere in the country.

The cultivated area within the GAEPS is at least as good an indicator of the impact of the scheme as the proportion of participating farmers. The GAEPS covers a higher proportion of the cultivated area than might be inferred from the participation rate. The GAEPS has almost total coverage (96%) in southern Finland and the participation of cultivated land in the whole country is as high as 90%. The original target of 87% coverage of the total arable land area was slightly exceeded.

More than 5000 farms participated in the SPS programme in 1995 and the majority of the contracts were made for 5 years. The SPS support totalled 80 million FIM in 1995, and most of it (30 million FIM) was allocated to organic production or organic conversion. Establishment of riparian zones, sedimentation ponds or wetlands was not as popular as expected, mainly because of the lack of time for preparing plans and applications. It is expected that more applications for implementing these measures will be presented in the future.

A posteriori monitoring

A monitoring programme covering participation and changes in agricultural practices, as well as impacts on environmental loadings, was launched at the outset of the programme (1995). Information on participation and certain preconditions, such as winter cover cropping, will be included in official statistics, while changes in most of the other practices, as well as environmental changes, are monitored by collecting data from selected areas. To achieve this, four study areas representing contrasting natural conditions and agricultural structure were selected (Fig. 13.3). The field parcel-based data from these areas were collected for 1994 (before Finland's accession to the EU) and 1995 (first year under the Common Agricultural Policy) by interviewing farmers. The next interview was undertaken in winter 1997/98 to cover the years 1996 and 1997 and for the third time in 2000, covering the years 1998 and 1999. In this way, changes in agricultural practices will be monitored.

The environmental assessments will be made using mathematical models to calculate individual loss estimates for all relevant climate–soil–crop–topography–practice combinations. The model estimates will then be combined with spatial data collected from the four selected study areas by a GIS technique. The initial results on changes in farm practices are presented here. Work is continuing to derive environmental loss data. Research on the impacts of the programme on habitat and species diversity was also launched in 1995. This programme consists of monitoring selected animal species on field-margins in selected areas, systematic grid-based and habitat-based plant species mapping, and landscape ecology assessment based on monitoring changes in habitat diversity.

When comparing fertilizer levels in 1995 with those of 1994, a slight decrease was detected for some of the most commonly cultivated crops. Stocking rates decreased, but the limit (1.5 LU ha^{-1}) was still exceeded on many farms. The clearest change was in winter green cover: it had significantly increased, particularly in southern Finland, where cereal production is concentrated. The winter green cover is partly achieved by reduced tillage techniques (leaving more residues).

One of the aims of the programme is to change the season for manure application from autumn/winter to the spring/summer period. In three of the

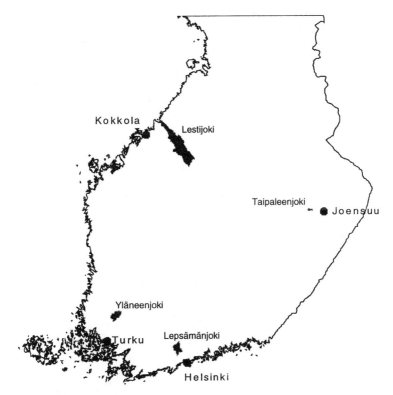

Fig. 13.3. Study areas selected for monitoring the Finnish Agri-Environmental Programme.

four study areas a slight change towards spring/summer was observed, but in one area the opposite occurred.

Nutrient Balances in the Study Areas

The nitrogen balances in the study areas show an increase in cumulative nitrogen surplus from 1994 to 1995 in the two southern study areas (Lepsämänjoki and Yläneenjoki), but a decrease in the other two areas (Fig. 13.4). Since the differences in nitrogen use (chemical fertilizers + manure) between these years were not significant, the most obvious reason for the changes in the surpluses is differences in nitrogen yields which were lower in 1995 in the two southern areas, while higher in the other two areas. This dependency on weather conditions, observed in the average national balances (see Table 13.1), must also be taken into account, particularly if incentives or levies are based on nutrient balances.

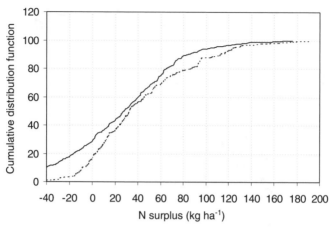

Lestijoki

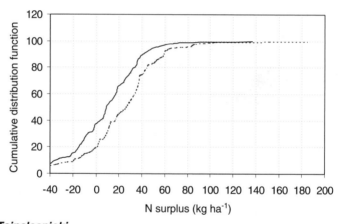

Taipaleenjoki

Fig. 13.4. Nitrogen surpluses in the four study areas in 1994 (-----) and 1995 (——).

The cumulative distributions from the study areas also show high variability between the areas. Taking 60 kg ha^{-1} as an example, the proportion of field parcels exceeding this limit varied from 4 to 30% in 1995. This is a clear indication of the different intensities of farming in different regions.

DISCUSSION

The nutrient balance calculations (NBCs) can be seen as a way of examining farm pollution (nutrient runoffs) from a point-source perspective. Thus, the

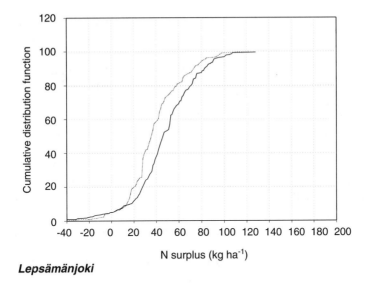

Lepsämänjoki

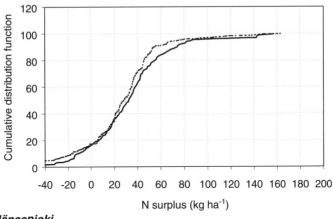

Yläneenjoki

Fig. 13.4. *contd.*

usual nonpoint context for agricultural pollution is circumvented, and conventional environmental control policies can be applied to the polluting discharges. As attractive as this seems, it would be unwise to make hasty assumptions about the feasibility of the NBCs in environmental control policies before also scrutinizing the problems associated with their derivation.

One major obstacle in using NBCs in agri-environmental policy is that they do not convey information about the fate of the nutrients leaving the agricultural system: NBCs merely state and quantify the situation but do not reveal whether the nutrients accumulate in the soil, leach to watercourses or volatilize into the air. Therefore, if a policy measure based on NBCs is set up, the specific environmental effects, e.g. with respect to the water quality, may remain unclear.

Perhaps even more importantly, the weather plays an important role in the agricultural production process, in particular in northern latitudes. Crop yields, and consequently nutrient surpluses, vary considerably in Finland because of the northern climate. For example, the year 1987 was very poor for crop growth in Finland and nutrient uptake was 30% lower than in the previous year, resulting in high nutrient surpluses.

In addition, the purpose for which the NBCs are used reflects the accuracy of the calculations. If farmers feel that NBCs will provide them with new and relevant information for improving the effectiveness of nutrient use, they will provide accurate information for the NBC calculations. But, if they feel that the NBCs will be used to penalize their farms, they may be tempted to give less accurate information.

The Finnish Agri-Environmental Programme (FAEP) is expected to considerably decrease nutrient runoffs from agriculture. It has been estimated that, in the long run, the adoption of the Agri-Environmental Support Scheme will decrease both erosion and phosphorus and nitrogen losses into watercourses by about 25–40%. Close to 90% of the arable land area is under the FAEP. The progress of the programme can be assessed with NBCs. However, the first year of implementing the FAEP in 1995 had a late start, and the results from that year show little change in nitrogen surpluses in the four research areas reported in this study. Consequently, calculation of NBCs for the coming years will provide a better indication of how well the goals of the programme may be achieved.

The FAEP presents several criteria for farmers to fulfil in order to be eligible for a compensatory payment. This is a very regulatory approach. An alternative policy approach would be to introduce environmental goals and the means for attaining these goals by providing the farmers with know-how on the ways to improve the environmental efficiency of their farms. NBCs provide one way in which this efficiency can be assessed both at the farm level and at a more general level. Thus, the NBC information serves both the farmer and administrators in assessing the environmental impacts of the programme. However, the limitations of the calculations must be recognized and taken into account.

The NBCs can be viewed as a way of simplifying the use of the command-and-control policy measures. Instead of setting up multiple regulations aiming at improving the environmental effects of farming practices, calculating the nutrient surpluses gives a simple assessment on the impact a farm has on the environment. Then, it is up to the farmer to decide what to do to ameliorate the nutrient use efficiency on his farm.

The NBCs can provide farmers and administrators with important information regarding the impact of agriculture on the environment. The accuracy and feasibility of this information is influenced by many different factors. These factors must be recognized when making the NBCs, and before using the information gained from them at farm level, or for agri-environmental policy formulation.

REFERENCES

Brouwer, F.M., Godeschalk, F.E., Hellegers, P.J.G.J. and Kelholt, H.J. (1995) *Mineral Balances at Farm Level in the European Union.* Agricultural Economics Research Institute (LEI-DLO), Onderzoekverslag 137, The Netherlands.

EEC (1992) Council Regulation (EEC) Nr. 2078/92 on agricultural production methods compatible with the requirements of the protection of the environment and the maintenance of the countryside. *Official Journal of the European Communities* L215, 85-90.

Helander, J. (1996) Sikataloustarkkailun tulokset 1995. Kannattavuus ja ravinteiden käyttö käsikynkässä. *Sika* 4, 7–9.

Lankoski, J. (1996) Agricultural pollution control through economic instruments based on mineral balances. *Maatalouden Taloudellisen Tutkimuslaitoksen Tiedonatoja* 205, 37–52.

Lord, E. and Anthony, S. (1997) Nutrient balances as environmental indicators for the UK. *Table document of Eurostat.* Doc. Agrienv/97/7.

Mukula, J. and Rantanen, O. (1989) Climatic risks to the yield and quality of field crops in Finland: III-VII. *Annales Agriculturae Fenniae* 28,1, Seria Agricultura 95, 1–43.

OECD (1997) Environmental Indicators for Agriculture. OECD, Paris.

Schleef, K.-H. and Kleinhanß, W. (1994) *Mineral Balances in Agriculture in the EU.* Part 1: *The Regional Level.* Bundesforschungsanstalt für Landwirtschaft, Braunschweig.

Teigen, L. D. and Thomas, M. (1995) *Weather and Yield, 1950–1993; Relationships, Distribution, and Data.* Commercial Agriculture Division, Economic Research Service, USDA.

Sustainability Indicators for Multiple Land Use in the Uplands

14

Bob Crabtree

INTRODUCTION

There is a high degree of consensus on what constitutes a useful framework for the development of *environmental* indicators. *State-of-the-environment* indicators characterize environmental condition, and repeated observation of the environmental state over time provides the basis for monitoring change. Other types of indicator can be added, depending on the context. Where the focus is primarily on environmental change and its causes, *pressure* and *stressor* indicators provide information on the underlying causes of environmental change. These were used in the Dobříš assessment which produced the set of European level environmental indicators (Stanners and Bourdeau, 1995). OECD (1994) incorporates a further set of *response* indicators in order to capture the policy response to environmental degradation and the feedback effects on environmental pressures. This *pressure–state–response* (PSR) framework has become widely accepted for developing national-level indicators and international comparison.

What constitute the defining characteristics of *sustainability* and *sustainable development* indicators are much less clear. The lack of consensus reflects a range of possible functions for such indicators: to raise awareness, to identify and measure sustainability (or unsustainability), or to monitor change and prompt a policy response. It also reflects differences in the definition and interpretation of sustainable development, on which a substantial amount of literature has been produced.

Heal (1996) defines the key sustainability concerns as:

©CAB INTERNATIONAL 1999. *Environmental Indicators and Agricultural Policy* (eds F.M. Brouwer and J.R. Crabtree)

- recognition of the long-run impact of resource and environmental constraints on patterns of development and consumption; and
- concern for the well-being of future generations, particularly in so far as this is affected by their access to natural resources and environmental goods.

While few would disagree with Heal, perspectives differ widely. Under the concept of *weak sustainability* (Solow, 1993), a sustainable path is one characterized by non-diminishing personal welfare from generation to generation. This is interpreted by Pearce *et al.* (1994) as a requirement to maintain the total stock of natural and man-made capital, but without any special conditions attached to natural assets. *Strong sustainability* places constraints on the allowable solution between natural and other forms of capital, giving rise to the concept of criticality in environmental stocks (Pearce and Atkinson, 1995). This emphasizes the issue of allowable trade-offs between and within capital stocks. Others have taken a more ecocentric and instrumental perspective which emphasizes the role of environmental protection, emission policies and use of non-renewables (Jacobs, 1991). In so far as there is consistency across perspectives, it relates to the long-term dynamics of change in the economy–environment interrelationship, and emphasizes protection of both economic functions and ecological systems and capabilities. Going beyond this, sustainability is evidently a spatial concept: spatial scales will differ (from global to local level) depending on environmental attribute, and localities will differ in the characteristics of their economies and environments. Both top-down and bottom-up responses occur, with participation of stakeholders at all levels emphasized under Chapter 28 of Agenda 21 (UNCED, 1992).

This paper examines the nature and role of local level indicators in sustainable development. It draws on case study experience in an environmentally sensitive upland/ mountainous area in the UK.

SUSTAINABILITY INDICATORS

We begin by taking the broad view that sustainability indicators are those useful for informing policy and resource management within a sustainability decision framework. Hence, the indicators will include many conventional economic and environmental state indicators. On that basis it is clear that *sustainability* indicators do not exist as a totally separate and distinguishable indicator set. Nevertheless, there are distinctive types of indicators particular to the issues involved in policy formulation for sustainable development and these are described below.

Macro and Sectoral Measures of Sustainable Development

One type of sustainability indicator attempts to identify, within a single measure, whether an economy or a sector is sustainable. These indicators generally take the form of modifications to national accounts which are known to indicate economic welfare inadequately and fail to account for depletion of environmental assets. Green net national product, which adjusts gross national product (GNP) for depreciation in man-made and natural capital, has been widely proposed as a sustainability indicator. Daly and Cobb (1989) define their index of sustainable economic welfare which modifies GNP for a number of environmental and other welfare changes. Pearce and Atkinson (1993, 1995) define a 'Z' indicator of weak sustainability and propose additional indicators of strong sustainability which vary in definition depending on the treatment of 'critical' natural capital.

Recent attempts have been made to produce sector level indicators (Vaze and Balchin, 1996; Whitby and Adger, 1996). Sector-level measures can give valuable insights for policy, particularly if disaggregated to different forms of environmental damage or investment. They are limited as stand-alone measures of sector sustainability because of interdependencies between the capital stocks of different sectors.

Pressure, State and Performance Indicators

Most of the indicators produced to inform on sustainable development are selected by context, reflecting issues of policy concern and particularly economy–environment interactions. Disaggregated, multifaceted approaches have been found to be more informative for policy development, with indicators selected in relation to their ability to inform on key dimensions of sustainability (Anderson, 1994; World Bank, 1995; DOE, 1996). These will principally be pressure and state indicators relating to processes and sectors in the economy with interlink to the environment either through activities (e.g. transport), resource demands (e.g. water) and environmental impacts (e.g. greenhouse gas emissions, acidification, and impacts on biodiversity and landscape).

The World Bank (1995) adopted such a disaggregated approach, linked to Agenda 21, and identified a series of environmental sources, sinks and life support functions, together with economic, social and institutional dimensions in sustainability. For each separate dimension, driving force, state and response indicators are presented in a matrix format. The UK selection of indicators (DOE, 1996) was based on a similarly disaggregated approach and used a modified OECD PSR framework. This gave greater prominence to economy–environment interactions and the role different actors play in responding to economic and environmental change.

Opschoor and Reijnders (1991) identified a specific role of sustainability indicators: to present environmental performance in the context of sustainability criteria by measuring the 'distance' between a threshold and the current environmental state. Thresholds are defined in relation to legal, scientific or policy-derived norms. Such performance indicators can be defined to capture concepts such as critical thresholds and excess, with their implications for environmental damage. Barbier (1989) has proposed conditions for sustainability that are consistent with a natural capital framework and can also be interpreted as thresholds:

- rates of utilization of renewable resources not greater than regeneration rates;
- rates of use of non-renewables not greater than the rate of creation of renewable substitutes; and
- waste emissions not greater than the assimilative capacities of the environment.

In such a context, performance indicators measure either the extent of excess or the margin of sustainability. Such indicators are only fully applicable where thresholds are defined and determine behaviour. In order for this to happen, a threshold has to be established through the socio-political process and ultimately represented as a legally defined maximum or as a binding policy target. The number of legally defined critical thresholds is increasing (for example, maximum levels of nitrate in drinking water, minimal damage to habitats and species protected under the Habitats Directive, 92/43/EEC), and this trend seems set to continue.

The concept of criticality (the condition for sustainability that the stock of critical natural capital must not decline) (Pearce and Atkinson, 1995) can also be interpreted in terms of thresholds, on which sustainability is conditional. However, there is little consensus on how to define critical capital, and how to incorporate reversible and irreversible loss and the risk to sustainability associated with the uncertainty attached to capital recovery. Where an underlying biophysical criticality (in terms of thresholds for human health or ecosystem function) can be identified, the setting of thresholds may be less ambiguous. In practice, however, attempts to produce a workable framework for natural resource management based on criticality have had only limited success (Gillespie and Shepherd, 1995).

As indicators become better developed, a particular role for sustainability indicators must be to interpret and inform on the underlying concepts of sustainability theory (e.g. intergenerational time-frame, criticality, spatial and temporal substitution of stocks, irreversibility, uncertainty). Ultimately the incorporation of these aspects of resource management should distinguish sustainability indicators more clearly from classical environmental performance indicators.

Policy Response Indicators

Paralleling the *response* element of the OECD PSR framework, there is a final set of indicators that monitors society's response to sustainability. Important here are defensive expenditures on the environment and measures to limit loss of natural resources through increased efficiency in resource use, recycling and the development of substitutes. While indicators have a role here, it may be a limited one. Policy responses typically involve the application of economic instruments or the setting of environmental standards. Such policy responses are best recorded not through quantitative monitoring of the indicator type, but through assessment of their cost and effectiveness. At the local level, an important role for indicators is the setting of targets that will lead to improved sustainability.

LOCAL SUSTAINABILITY INDICATORS

Although environment–economy interactions clearly occur at local level, many of the pressures and impacts spread across local boundaries and are not under the control of local policies. As the spatial scale is reduced below the global and national levels, so the scope for measuring sustainability or even interpreting change in a sustainability context becomes much reduced. Figure 14.1 illustrates the interpretation problems that occur when an area is considered as an entity separate from the wider economy and environment. It is uncommon for local economies and markets, with the possible exception of local labour markets, to be highly differentiated by transfer costs or differences in consumer tastes. This interconnectedness with the wider economy carries

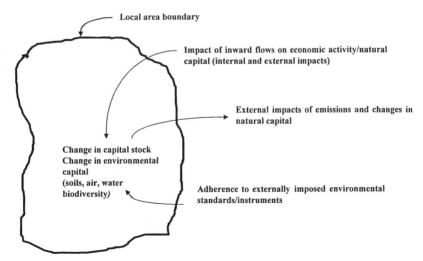

Fig. 14.1. Transboundary effects and local area sustainability.

the implied sustainability impacts to a larger spatial scale. Similarly, environmental change both within and external to a local area may reflect transboundary fluxes. Numerous environmental impacts are not locally contained or may be difficult to interpret at a local scale (e.g. greenhouse gas and acid emissions; biodiversity impacts). Table 14.1 indicates appropriate scales for agricultural emissions and impacts on environmental capital.

Against this background, local sustainability indicators will be most useful where:

• they inform local actors and local policy-making; and
• economy–environment impacts are firmly located in local space and important for local welfare.

Local indicators will be concentrated on local issues which can be addressed through local policy instruments. The indicators developed in the Local Agenda 21 context in the UK cover a wide range of issues, including social welfare, waste treatment, transport and urban environmental improvement (LGMB, 1995). However, where indicators of sustainable development become partial and fragmented, and this will be the case when developed locally, there is a danger that any underlying conceptual basis is lacking: indicators then lose any coherence. UK local government experience is that when community input into the development of indicators occurs 'the process of developing and using sustainability indicators is an evolutionary one' and that there 'can be no agreed pattern or template for the process' (LGMB, 1995). Sustainability indicators can then become hard to distinguish from environmental or local authority performance indicators.

SUSTAINABLE MANAGEMENT OF THE CAIRNGORMS MOUNTAIN AREA

The remainder of this chapter presents a case study on the development of local sustainability indicators for the Cairngorms (Fig. 14.2), a mountainous area of

Table 14.1. Appropriate spatial scales for environmental dimensions.

Environmental dimension	Emission/damage (examples)	Spatial scale
Soil	Heavy metals	Site
Air	Acidification	Local → international
	Greenhouse gases	Local → international → global
Surface water	Nitrogen	Site → catchment → international
	Phosphorus	Site → catchment
Biodiversity	Numerous	Local → ecosystems → biosphere
Specific habitats and species	Numerous	Site → local
Landscape	Numerous	Site → local → cultural landscape

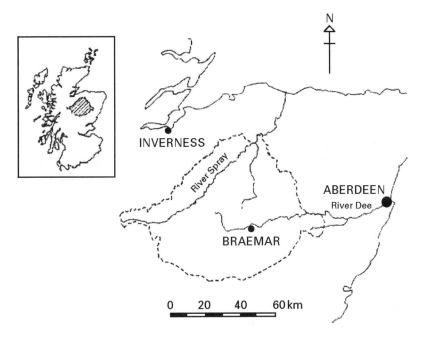

Fig. 14.2. Location of the Cairngorms Partnership area in NE Scotland.

519 000 ha in the north-east of Scotland. It is an area of outstanding environmental value with substantial areas of mountains, moorland, forests and wetland (Nethersole-Thomson and Watson, 1974; Conroy *et al.*, 1990). Seventeen per cent of the area is proposed for designation under the Habitats Directive of the EU (92/43/EEC). Land use encompasses agriculture, forestry, and deer and grouse shooting. The area is valued for tourism and recreation, and includes three ski resorts, development of the latter being possibly one of the most contentious environmental issues in Scotland in recent years. The area has a population of slightly over 17 000 and provides employment (employees and self-employed) for around 8000 (NOMIS, 1991).

In 1994, the UK government created the Cairngorms Partnership (Scottish Office, 1994) with a policy remit to coordinate the activities of existing local and regional authorities, government bodies and interested groups in the Cairngorms and facilitate sustainable management of the area, taking account of local economic, environmental and community concerns. The main guiding 'principle' was that of sustainability, interpreted as safeguarding the environment while promoting sustainable economic growth in the area. However, the principle was not translated by government into a decision framework which could guide the crucial trade-off decisions between environmental protection and economic and leisure activities that form the main policy challenge. No additional powers were granted, apart from a limited budget of £0.5 million

per year, the emphasis being on voluntary action, persuasion and collabora-
tive activity with other stakeholders.

A Framework for the Development of Indicators

The Partnership required the development of a set of key indicators that would
assist in the design of a strategy for sustainable management and the monitoring of their success in meeting the objectives set by government. In terms of the
types of sustainability indicators identified previously (see above), the single
macro indicator approach based on accounting methods was regarded as irrel-
evant at the local scale. What was required was a series of disaggregated indicators to inform policy on specific themes. The aim was to use a modified PSR
approach (Fig. 14.3) and to concentrate on pressure and state indicators,
(including performance indicators where they could be defined) and policy-
response indicators.

   Several themes were identified. The environmental emphasis was on those
dimensions relevant at a 'site' or 'local' scale: soils, surface water, biodiversity,
specific habitats, species and landscape (see Table 14.1). Six social and economic themes relevant to the local area were included, together with a number
of institutional themes (planning and transport, environmental education and
nature conservation). The themes provided a framework in which the current
state could be defined and the processes determining economic and environmental change examined. They were selected as relevant to the policy remit
and competence of the Partnership, and concentrated on local environmental
and cultural capital, together with those elements of the local economy and
structure that had some measure of local policy input. These included local
housing and locally prescribed initiatives for rural development. The study was
undertaken by a group of experts who relied on a mixture of previous personal
experience, published information and databases, and interviews with key
informants.

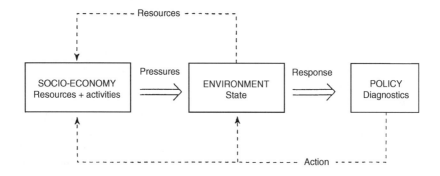

Fig. 14.3.   The conceptual framework for sustainability indicators.

Socio-economic Activity and the Processes of Environmental and Institutional Change

Socio-economy

The socio-economic themes covered the economy, economic and social development, housing and settlements, tourism, recreation and agriculture. Analysis of the local economy revealed that its structure and performance reflected its remote location, generally poorly skilled labour force and lack of competitivity in most sectors other than recreation and tourism. The labour market was poorly structured with a predominance of part-time work and the economy lacked diversity, although the priorities of local government and development agencies were to stimulate economic activity through support for training, investment and marketing. One of the constraints on economic development and the retention of local labour was an under-supply of low-cost housing due to competition from in-migrants.

The activities in the area which generate significant environmental impacts are recreation and tourism, agriculture, forestry and estate management. Taking these in turn, tourism and recreation are the major economic activities in the area (41% of all employees in the 1991 Population Census were employed in the catering and distribution sectors). The environmental impacts are mainly negative. Table 14.2 refers to tourism impacts and provides

Table 14.2. Environmental impacts of major tourist facilities and activities in the Cairngorms Area.

	Impact	Associated environmental impacts
Facility		
Ski resorts	Negative	Erosion; visual intrusion; disturbance to wildlife; changes to site hydrology
Aviemore Mountain Resort (principal tourist centre)	Negative	Visual intrusion
Popular tourism routes	Negative	Facilitate tourist access to sensitive areas
Larger hotels and timeshare resorts	Negative	Large volumes of trade waste produced requiring disposal
Activities		
Car driving (over 80% of tourists in the area travel by car)	Negative	Congestion; air pollution; road-side erosion
Walking/hiking/countryside activities	Negative	Erosion; wildlife disturbance
Off-road and four-wheel driving	Negative	Bare ground; erosion; wildlife disturbance
Mountain biking	Negative	Erosion, damage to paths and vegetation

an example of the diversity of processes by which economic activity is linked to environmental change.

Agriculture is important both to the economy and the environment in the Cairngorms. It accounts for 8% of employment and takes place on 74% of the land. The main negative environmental impacts reflect high stocking levels, drainage and intensification. In the longer term, strong economic and social factors inhibiting succession on hill farms (Shucksmith and Smith, 1991) are leading to gradual changes in land tenure and farm amalgamation. Such structural change can be associated with intensification and losses in habitat diversity (Munton and Marsdon, 1991; Marsdon *et al.*, 1993) although increased afforestation of land may also occur. However, the operation of the agri-environmental measures under EC Regulation 2078/92 is expected to lead to increased protection and enhancement of a range of habitats including herb-rich grassland, moorland, native woodland and wetlands.

A number of other important effects of human activity in the area were identified but these had origins largely exogenous to the area (e.g. acid deposition, climate change) or were not clearly related to economic or social activity (e.g. fires).

Environment

The environmental themes were first examined by assessing their condition and identifying the principal pressures from human activity (Table 14.3).

Condition varied greatly both within and between features (Table 14.3). The sites in best condition were sometimes those under statutory protection or remote from disturbance, but many of the causes of concern such as grazing,

Table 14.3.   Condition of environmental capital and major pressures.

Theme	State	Pressures
Woodlands	Poor–good	Grazing, fires, fragmentation of native forest
Water and fisheries	Moderate–good	Algal blooms, effluents, climatic change, water management, tourism
Wetlands	Poor–good	Grazing, harvesting foliage, pollution, drainage, climatic change
Moorlands	Poor–good	Grazing, burning, peat erosion
Mountains	Poor–good	Grazing, acid deposition, construction, fragmentation, climatic change
Gamebirds	Poor–declining	Habitat loss, disease, fences, pollution
Red deer	Poor–moderate	Grazing, habitat deterioration
Landscape	Moderate–good	Numerous
Cultural heritage	Poor–good	Ignorance, afforestation, construction work, drainage, recreation

pollution, fires and climatic change were pervasive. Exceptions were recreation and construction impacts which affected only a small but conspicuous proportion of the area. A frequently identified concern was grazing, which affected most of the ecological features; habitat change or deterioration in various forms was found to be widespread. Other concerns affecting more than one feature were pollution, acid deposition and climatic change. In some cases there were complex ecosystem effects such as habitat damage caused by high deer populations, where there was no direct linkage with local economic activity or development. This apart, virtually all the pressures on the environment were the result of human (generally economic) activity, within the local area and thus potentially open to local intervention. The principal exceptions were acidification and climate change which reflected impacts of economic activity at the national or global level.

Institutions

The institutional response to the issue of sustainability in the Cairngorms has been substantial. The existence of the Partnership itself reflects the strengthening of policy involvement in the area. This apart, the environmental instruments of land purchase, designation under national and EU Directives and management agreements with farmers and landowners were widely applied. In the socio-economic domain, all of the area was designated Objective 1 or 5b under EU Structural policy (Council Regulation, 2081/93) (with associated LEADER programmes) and received support from Local Enterprise Companies.

Indicators and Interpretation

Several criteria were used to select indicators for monitoring and management. Operational feasibility was clearly important and it was found impossible to develop many of the pressures into useful indicators because of problems of measurement. Where indicators were not ruled out on feasibility grounds, an informal cost–benefit framework was used to rank indicators. Indicators that were available without additional cost, apart from the administrative cost of extraction, and which would usefully inform policy, were given a high priority. The more difficult decisions related to indicators not or only partially available, where additional costs would be incurred in developing the indicator. Indicators were only proposed if the cost was not likely to be excessive in relation to the benefits from improvement to the information base of the Partnership. Many of the economic indicators required the reformulation of existing census and survey data. In this case a cost-effective strategy was to liaise with government and other agencies in an attempt to obtain data sets in a more relevant form. Current environmental monitoring fell well short of that ideally required for local management, but the cost of establishing new indicators was in many cases prohibitive.

Table 14.4 gives the number of indicators of each type proposed for each theme and gives two examples of indicators for each category. Overall, a substantial number of indicators were proposed of which several are based on currently available data sets. Others are available for only part of the area, or have not been recorded consistently over time. Some are not being recorded or are unavailable because of confidentiality. In general, more deficiencies were identified in natural and physical history aspects than in social, economic and cultural history ones, where a relatively high proportion of indicator data was already being recorded. This has been driven by the practical needs of planners, developers and local and central government, so that data for the majority of indicators are already in the public domain.

As might be expected, most of the pressure indicators are located under the social, economic and institutional themes. Pressure indicators developed under the environmental themes tended to be specific to those themes, but in most cases there were more pervasive influences from development activity not separately identified for each environmental theme. The environmental and cultural indicators concentrated on measuring the state of the environmental resources. The policy indicators fell into two main categories: those designed to stimulate and control development activities and those designed for environmental protection and enhancement. Together, these make up a sizeable set, reflecting the wide range of policy intervention in the area. Unlike the environmental indicators, where their introduction requires a system of measurement and monitoring, the policy indicators require the development of interfaces with appropriate institutions in order to monitor policy impacts and influence policy development.

Considerable difficulty was found in defining *performance* indicators, despite their obvious value for sustainable management. In most cases it was not possible to define capacity thresholds because of a lack of:

- experience of individual indicators and their interpretation;
- nationally agreed standards of most types of indicators; and
- consensus mechanisms for setting sustainability standards.

Over time, however, improvement in the indicator set coupled with recorded changes in state indicators could well provide the stimulus to define thresholds as targets for sustainable management.

SUSTAINABILITY ASSESSMENT

The majority of the proposed indicators have as their main function that of informing the local policy process. They do not, with a few exceptions, indicate whether current economic activity or the current supply of environmental goods and services are sustainable over the longer term. In fact, few economic

Table 14.4. Thematic indicator matrix.

Theme	Socio-economic pressure indicators		Environmental state indicators		Policy indicators	
	No.	Description	No.	Description	No.	Description
Environment and cultural						
Woodlands	1	Standing stock of trees	8	Structural and spatial diversity	1	Government afforestation incentives
				Landscape character assessment		
Water and fisheries	1	Effluents	19	Headwater acidification		
				Loch eutrophication		
Wetlands	6	Areas drained or reclaimed	7	Bird numbers at important sites	1	Impact of ESA management
		Levels of recreational use		Nutrient status of lochs and rivers		
Moorlands	1	Herbivore grazing	6	Condition of base-rich grassland		
				Condition of blanket bog		
Mountains	12	Anthropogenic pollution of rivers	4	Extent of bare ground		
		Visitor numbers		Bird numbers		
Gamebirds			5	Species populations		
				Habitat condition		
Red Deer			7	Species populations		
				Habitat condition		
Landscape			4	Extent of tranquil areas		
				Visual intrusions in the montane zone		
Cultural heritage	9	Village development			1	Archive quality

Continued over

Table 14.4. *contd.*

Theme	Socio-economic pressure indicators		Environmental state indicators		Policy indicators	
	No.	Description	No.	Description	No.	Description
Social and economic						
Economy	5	Population change Employment change				
Economic development	10	Planning applications New business starts			2	Planning policies; activities of economic development and community bodies
Housing and settlements	12	Migration rates House completions				
Tourism	7	Stock of accommodation Occupancy rates for accommodation				
Outdoor recreation	5	Number of people involved Distribution of walkers	2	Bare ground on paths Length and extent of paths	1	Access agreements
Agriculture	3	Impacts of farming change Land and labour use in agriculture			2	ESA[a] scheme uptake ESA scheme monitoring
Institutional						
Planning and transport	13	Change in traffic flows Population change			4	Sustainable transport policies Alternative modes of transport
Environmental education			8	Number and level of courses offered; number of interpretative displays		
Nature conservation	12	Damaging actions on protected sites; number and types of planning applications	8	Habitat condition Footpath condition	2	Protected area policies ESA monitoring

[a] ESA refers to the Environmentally Sensitive Area designated under 2078/92.

or social indicators can be interpreted unambiguously in a local sustainability context. This is only possible where there are no negative environmental impacts or where the impacts are below any possible critical thresholds. For example, with the indicators for tourist activity, increased tourism supports local employment and incomes but may aggravate pressures on the carrying capacity of the mountain area (Table 14.2). The sustainability interpretation of a tourism indicator is thus ambiguous. Similarly, income pressure on agriculture may lead to less intensive or alternative land uses (e.g. afforestation) where the environmental impacts may be difficult to interpret. It became apparent that in such circumstances indicators needed to be supplemented with an understanding of the dynamics of process if they were to be fully informative for policy. Only then can assessment be made of the case for policy intervention.

Certain indicators could be interpreted directly as indicating unsustainability if the tenets of strong sustainability are assumed and all impacts are contained within the local area. With these assumptions, unsustainability occurs where, for example:

- waste emissions exceed, or are predicted to exceed, environmental capacity;
- other environmental impacts (e.g. recreation) exceed, or are predicted to exceed, environmental capacity;
- extent and quality of specified habitats are in decline;
- there is a deterioration in key socio-economic indicators, measuring relative performance (as defined, for example, by Copus and Crabtree, 1996).

Judged against these criteria there was evidence of unsustainability, although this was often localized or reflecting pressures arising externally to the area and not subject to local management. For example, pollution impacts were noted on water resources and gamebirds but of the causative agents only nitrate and phosphorus pollution were locally generated. The indicators suggested that environmental capacity was being exceeded. There was strong evidence that climatic change was exceeding the capacity of the environment to adapt but the source of this pressure is primarily external. The major internal pressures that raised questions about sustainability were grazing levels, recreation activity and commercial management of forest and mountain areas leading to habitat fragmentation.

CONCLUSIONS

Evidence from the case study indicates that the main function of the indicators was to inform policy-makers and other stakeholders about the state of the local economy and environment, and the pressures driving change. Using indicators to measure sustainability directly at local level proved problematic, although this was not the prime aim in the study. Not only were the environmental data generally inadequate, despite extensive previous research in the area, but criteria on which to base local criticality were not available. Nevertheless, some pointers on sustainability and unsustainability were forthcoming. These were most clear-cut where the environmental stock was deteriorating due to locally derived pressures. Where external pressures were involved (acidification, climate change) the scope for local action was weak and the relevance of the indicators to local policy correspondingly limited.

The difficulty in defining performance indicators in relation to critical thresholds reflected the fact that there were no existing thresholds; for policy purposes they could only be established through the mechanism of negotiation. Over time, greater directionality in local sustainability policy will develop. Experience suggests that a variety of evidence will contribute to the process. For example, with standing waters, hindcasting evidence of pre-industrial (pristine) conditions is being used as one baseline for targeting water quality improvement (Fozzard *et al.*, 1997). Current policy for forestry in the Cairngorms is to increase the area of native pine substantially, given the losses in woodland cover that have occurred over time (Hester *et al.*, 1996). However, the ability to define policy goals and to incorporate sustainability concepts such as criticality and economy–environmental trade-offs will depend critically on implications for the different interests involved. To take the example of native pinewood expansion, there has been a common interest between environmental groups, landowners and local communities. All see some local benefit from planting under a government-financed incentive scheme for landowners. Such common interest is less obvious where environmentally driven constraints are proposed for tourist or recreational activity, and this severely limits the ability to agree on a local interpretation of sustainability.

More generally, indicators constitute a set of essential economic and environmental data to support policy formulation and delivery. They do, however, have limitations. Important determinants of local development are difficult to measure and capture as indicators (e.g. human capital, institutional capacity). Similarly, indicators of environmental change will be biased in relation to their feasibility of measurement. Over and above these limitations, successful local policy design requires a deeper understanding of the underlying biophysical, economic and policy processes than that provided by indicators alone. In the sustainable development context this means a detailed knowledge of the impacts of specific forms of economic development on environmental capital, and the local economic costs of environmental protection. Only then can policy be adequately informed to make decisions involving environmental trade-offs. This understanding provides the basis for influencing the activities of other agencies (for local development and environmental protection) in the interests of sustainable management.

Ultimately, sustainability indicators will be as much shaped by the policy process as developed to inform the process. That is, the selection of indicators reflects the interests involved and indicators may be rejected if they reflect past policy failures. Just as it has been argued that policy evaluation is an essentially political process (Palumbo, 1987) so the selection and use of indicators involves a political dimension. Local indicator development then reflects not only data availability and the extent to which externally defined critical thresholds exist, but the nature of representation and interest in the policy process itself.

REFERENCES

Anderson, V. (1994) *Alternative Economic Indicators*. Routledge, London.

Barbier, E.B. (1989) *Economics, Natural-Resource Scarcity and Development; Conventional and Alternative Views*. Earthscan, London.

Conroy, J.W.H., Watson, A. and Gunsson, A.R. (1990) *Caring for the High Mountains: Conservation of the Cairngorms*. Centre for Scottish Studies, University of Aberdeen.

Copus, A. and Crabtree, J.R. (1996) Indicators of socio-economic sustainability: an application to remote rural Scotland. *Journal of Rural Studies* 12, 41–54.

Daly, H.E. and Cobb, L. (1989) *For the Common Good*. Beacon Press, Boston.

DOE (1996) *Indicators of Sustainable Development for the United Kingdom*. Department of the Environment, HMSO, London.

Fozzard, I.R., Doughty, C.R. and Leatherland, T.M. (1997) Defining the quality of Scottish freshwater lochs. In: Boon, P.J. and Howell, D.C. (eds) *Freshwater Quality: Defining the Indefinable?* HMSO, London.

Gillespie, J. and Shepherd, P. (1995) *Establishing Criteria for Identifying Critical Natural Capital in the Terrestrial Environment*. English Nature Research Report 141, English Nature, Peterborough.

Heal, G.M. (1996) Interpreting sustainability. Paper given to the European Association of Environmental and Resource Economists, Lisbon, 1996.

Hester, A., Miller, D. and Towers, W. (1996) Landscape scale vegetation change in the Cairngorms, Scotland, 1946–1988, implications for land management. *Biological Conservation* 77, 41–51.

Jacobs, M. (1991) *The Green Economy: Environment Sustainable Development and the Politics of the Future*. Pluto, London.

LGMB (1995) *Sustainability Indicators Research Project: Consultants' Report of the Pilot Phase*. The Local Government Management Board, Luton.

Marsdon, T., Murdoch, J., Lowe, P., Munton, R. and Flynn, A. (1993) *Constructing the Countryside*. UCL Press, London.

Munton, R.J.C. and Marsdon, T.K. (1991) Occupancy change and the farmed landscape. *Environment and Planning* A23, 499–510.

Nethersole-Thomson, D. and Watson, A. (1974) *The Cairngorms*. Collins, London.

NOMIS (1991) *National On-line Manpower Information System*. University of Durham.

OECD (1994) *Environmental Indicators: OECD Core Set*. Organisation for Economic Co-operation and Development, Paris.

Opschoor, H. and Reijnders, L. (1991) Towards sustainable development indicators. In Kuik, O. and Verbruggen, H. (eds) *In Search of Indicators of Sustainable Development*. Kluwer Academic Press, Dordrecht.

Palumbo, D.J. (1987) Politics and evaluation. In: Palumbo, D.J. (ed.) *The Politics of Program Evaluation*. Sage, Newbury Park.

Pearce, D.W. and Atkinson, G.D. (1993). Capital theory and the measurement of sustainable development: an indicator of 'weak' sustainability. *Ecological Economics* 8, 103–108.

Pearce, D.W. and Atkinson, G.D. (1995) Measuring sustainable development. In: Bromley, D.W. (ed.) *Handbook of Environmental Economics*. Blackwell, Oxford.

Pearce, D.W., Atkinson, G.D. and Dubourg, W.R. (1994) The economics of sustainable development. *Annual Review of Energy Environment* 19, 457–474.

Scottish Office (1994) *Cairngorms Partnership*. Scottish Office, Edinburgh.

Shucksmith, D.M. and Smith, R. (1991) Farm household strategies and pluriactivity in upland Scotland. *Journal of Agricultural Economics* 42, 340–353.

Solow, R.M. (1993) Sustainability: an economist's perspective. In: Dorfman, R. and Dorman, N. (eds) *Economics of the Environment*, 3rd edn. Norton, New York.

Stanners, D. and Bourdeau, P. (1995) *Europe's Environment: the Dobrís Assessment*. European Environment Agency, Copenhagen.

UNCED (1992) *Agenda 21*. United Nations Conference on Environment and Development, Geneva.

Vaze, P. and Balchin, S. (1996) *The Pilot United Kingdom Environmental Accounts*. Office for National Statistics, London.

Whitby, M. and Adger, W.N. (1996) Natural and reproducible capital and the sustainability of land use in the UK. *Journal of Agricultural Economics* 47, 50–65.

World Bank (1995). *Monitoring Environmental Progress. A Report on Work in Progress.* The World Bank, Washington DC.

Implementation of Environmental Indicators in Policy Information Systems in Germany

15

Markus Meudt

INTRODUCTION

In the context of agricultural and environmental policy, information is required for problem identification, policy formulation, policy and programme implementation, monitoring and consistent evaluation. This need creates the demand for a tool that allows the detailed analysis of agricultural production and the environment, as well as the interaction between them. The Regionalized Agricultural and Environmental Modelling System (RAUMIS) was developed in the late 1980s as such a tool for integrated agricultural and environmental policy analysis.

RAUMIS was designed for monitoring, forecasting and policy simulation. The idea was to create a model for agricultural administration purposes and to promote dialogue with policy-makers. To be successful, the initial requirement was for a consistent database to be set up and for the model to be *highly detailed* (activity-based approach), so that account could be taken of individual variables relating to policy objectives and instruments, *transparent*, so that policy-makers could follow calculations and the data flow, and *up to date* and *flexible*, so that the latest data could be used in reference runs and simulations. Above all, the model had to have sound forecasting qualities, so that it could not only explain basic links, but would also provide highly accurate numerical forecasts for the most important variables relating to policy objectives.

This chapter describes briefly the methodological basis and structure of RAUMIS and includes detailed information on the environmental component of the model. Results are presented and discussed regarding an application for the target year 2005, in which the effects on agriculture and the environment of a status quo Common Agricultural Policy (CAP) (no change in policy measures, reference run) are compared with a free-trade-orientated and liberalized CAP.

METHODOLOGICAL BASIS AND STRUCTURE OF RAUMIS

An important feature of the RAUMIS concept is its activity-based approach. The agricultural sector of Germany is differentiated according to output and input groups following the principles of national accounts and the Economic Account for Agriculture (EAA). Production activities in RAUMIS are based on those of the EAA. The activity-based approach allows the differentiated analysis of the characteristic features of agricultural production, and interdependencies between the different production activities and the environment. The sectoral accounting framework ensures consistency with respect to physical and monetary flows. Figure 15.1 illustrates the gross production concept by showing an agricultural production account drawn up in that form.

Overall Structure of RAUMIS

RAUMIS is a mathematical programming model. More specifically it is a regionalized supply model (based on 431 regions) that represents the German agricultural sector, with 77 crop production activities (including set-aside programmes and 46 sub-activities with alternative technologies for less intensive production) and 16 animal production activities. The data used are consistent with the agricultural table of account. The most important characteristic of the model is a non-linear programming (NLP) approach with linear restrictions. The objective function of the NLP maximizes net income.

The model can be divided into a base model and a simulation model. The *base model* is used as a data-holding system for model calibration in a base-year run and for ex-post analysis as well as problem identification in the base years. It contains the fully quantified activity-based matrix for each model region. (There are existing base-year matrices for the counties of the former Federal Republic of Germany for the years 1979, 1983 and 1987; those for 1991 (and 1995 forthcoming) are quantified for all 431 German counties.) The RAUMIS *simulation model* focuses on the impact analysis of alternative agricultural and environmental policy measures in a target year. In simulation runs for the target year, consequences regarding production structure, input use, income and environmental impacts of these alternative measures are compared to a baseline scenario (reference run).

An important merit of the way the RAUMIS policy information system is used is the cooperation between the modelling group and the policy-maker in the definition of the relevant agricultural and environmental policy scenarios (Fig. 15.2). The definition could include the vector of output and input prices, yield assumption and scenario-specific parameters, e.g. nitrogen tax.

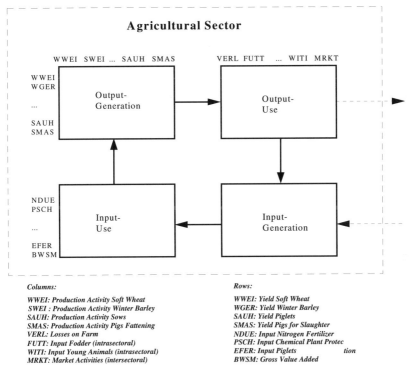

Fig. 15.1. Diagram of an activity-based table of account.

RAUMIS is characterized by its modular construction, shown in Fig. 15.2. It consists of three main modules (projection module, intensity module, non-linear programming approach). The variables resulting from the simulation module are the basis for a deeply differentiated policy impact analysis.

The first step in a simulation run is the *projection* of exogenous variables from the base year to the target year. Most of the variables, such as yield and variable inputs, are trend-projected, but it is also possible to overlay the trend projection by expert estimation. The projected variables can be divided into two groups. The first comprises *output coefficients*, most of which are trend projected. The regional values are mostly projected by sectoral growth rates of the European Union's SPEL (Sectoral Production and Income Model for Agriculture) system. The activity levels, crop output coefficients and milk output per cow have regionalized growth rates (based on ex-post data) which are bounded by an exogenously determined corridor of sectoral growth rates. The second group consists of input variables such as nutrient *input coefficients* for crop and animal production as well as physical coefficients which are dependent only on output. The price indices come from the scenario specification or the trend projections. A second group of input variables comprises input coefficients for energy, repair costs, labour and capital. These coefficients are not physical but monetary values. They are dependent on technology which, in

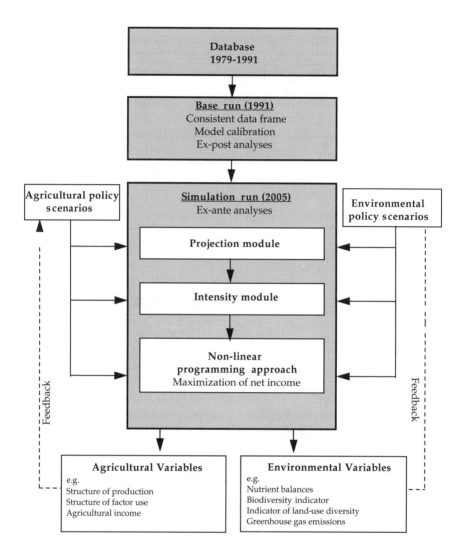

Fig. 15.2. Structure of the RAUMIS agricultural and environmental information system.

turn, is determined by farm size. The quality of land, as well as prices which are not given by experts in the scenario specification, are trend-projected. In the case of cereals, oilseeds, pulses, potatoes and sugar beet, the model uses time- and output-dependent quadratic cost functions (based on time series of representative farm data) to determine input coefficients for chemical plant protection. For seed input coefficients, the model uses linear functions. The coefficients for energy and repair costs depend on bundles of technologies, which are a function of farm size (Löhe, 1998).

The simultaneous modelling approach considers the effects of production and input price changes on crop production intensity and the degree of manure use. Such relationships are part of the *intensity module*. This module exists of two elements: a *yield function module* (YFM) and *manure technology module* (MTM).

In the YFM, individual yield functions for ten different crops were estimated via ordinary least-squares (OLS), based on field research data on nitrogen (N)-yield relationships. In order to abstract from different locations of the individual experiments, the maximum yield at each location was normalized to unity. Each series was standardized by dividing each N input by the N level corresponding to the maximum yield; at the same time, each yield of the series was divided by the maximum yield of the series.

The two main assumptions were:

- that the relation between N input and maximum yield remains consistent over time; and
- yield without fertilizer use is 20–30% lower in the medium to long term than in the short term. The latter assumption was made because the data for this approach were derived from field experiments in which the experimental fields were changed annually. This led to an overestimation of yield under zero fertilization.

The derivation of non-standardized region-specific yield functions from estimated standardized functions is carried out under the assumption that the amount of N under existing factor–product–price relations is optimal and that the relationship between yield level without fertilizer and maximum yield is the same in standardized and non-standardized functions.

To consider interactions between N input on the one hand, and the use of chemical plant protection (CPP), base fertilizer application and machine costs on the other, the product price is reduced by the amount of variable costs necessary for one yield unit (Fig. 15.3).

The MTM is used to determine endogenously the optimal unit capacity factor of manure N (differentiated into manure from cattle, pigs and poultry), dependent on the price of mineral fertilizer. Therefore, cost functions of four different manure technologies and different storage capacities were estimated for each of the three types of manure, dependent on the different manure unit capacity factors. Costs increase more than proportionately with an increase in the manure unit capacity factor.

Cost functions are regionalized with regard to the relation of total amount of N manure through N input needed by crops in this region. The regionally optimal unit capacity factors of each of the three manures can be derived based on these functions. The manure technology (unit capacity factor) is optimal, when the marginal costs of manure application are equal to the price of mineral N fertilizer.

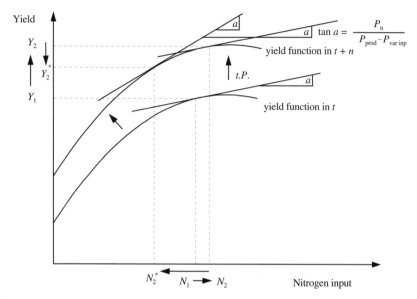

Fig. 15.3.   Intensity module of RAUMIS.

The linkage between YFM and MTM results from the aggregated marginal yield function which is derived from the yield function of each RAUMIS crop activity, which is similar to the regional N-demand function. The regional supply function from manure is derived from the manure technology function. If the usable N amount in a region exceeds the N needed by crops (caused by a special N price) the algorithm of the intensity module reduces the N price as long as 80% of the N needed by plants is provided by N from organic manure.

In a simulation run, the optimal N fertilizer input, yield and optimal variable input (basic fertilizer, energy and repair costs, chemical plant protection, labour costs) for each region and the main crop activities of the model were determined in the intensity module.

The trend-projected yields and N input combined with a vector of trend-projected prices are the basis for the intensity model. These data provoke a shift of the yield functions (technical progress) in Fig. 15.3. At N_2 the marginal productivity of N is equal to the relation between factor and product price. The result is a new yield, Y_2. Figure 15.3 shows the effect of product price reduction. As a result of this, a new optimum with lower input intensity will be realized: N_2^*, Y_2^*.

As mentioned above, the most important aspect of RAUMIS is the *non-linear programming* approach. A main feature of programming models such as RAUMIS is the problem of overspecialization. Because of a wide diversity of production activities and few empirical constraints on the regional level (NUTS 3) the optimization procedure reduces the realized number of activities. The traditional approach is to use constraints for calibrating the model results on observed production patterns in the base period. However, for policy simulations, such a set of constraints might be inappropriate. During the development of a new version of the RAUMIS model a *positive quadratic programming* approach (Bauer and Kasnakoglu, 1990; Howitt, 1995) was tested in an attempt to overcome this problem with overspecialization (Cypris, 1998).

The RAUMIS Environmental Modules

All environmental indicators in RAUMIS depend on the endogenously calculated production structure of German agriculture. The production structure in simulation runs results from the allocation mechanism of the model. The main objective of the environmental indicators is to show the direction and order of magnitude of environmental impacts, rather than to analyse specific local effects, and they therefore should not be seen as substitutes for detailed agricultural and environmental field studies. Rather, it is the intention that the RAUMIS environmental indicators should complement conventional environmental appraisal by giving a crude indication of potential endangered regions in ex-post analyses, and simulate impacts of agricultural and environmental policies on the environment.

Nutrient balances

RAUMIS is equipped with nutrient balances for nitrogen, phosphorus and potash. The balances are activity-based and are consistent for each region in the base year. The nitrogen balance has been the most intensively used environmental part of the model (Weingarten, 1996): hence the methodological concept of the implemented N balance is used here as an example of the conceptual development of all nutrient balances.

NITROGEN BALANCE. The input side of the N balance accounts for chemical fertilizer, manure, symbiotic and asymbiotic N_2 fixation and atmospheric inputs. N removal by crop harvesting and ammonia emissions is calculated endogenously. The N balance shows the amount of N that will be denitrified, leached out into the groundwater or accumulated in the soil. The accumulated net surplus can be interpreted as an indicator of 'potential pollution' (Weingarten, 1996)

Figure 15.4 shows the components of the N balance in RAUMIS. N balance is calculated for ex-post analyses as well as for ex-ante simulations. The N balance is calculated for each of the 431 model regions. Data referring to *chemical fertilizer input* are available only at the sectoral level for Germany and the regionalized input of chemical fertilizer is calculated as follows:

$$\text{Input}_{cf} = N_{req} - N_{man}$$

where:

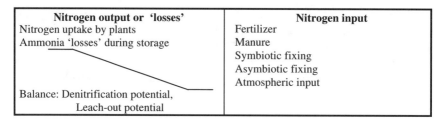

Nitrogen output or 'losses'	Nitrogen input
Nitrogen uptake by plants	Fertilizer
Ammonia 'losses' during storage	Manure
	Symbiotic fixing
	Asymbiotic fixing
	Atmospheric input
Balance: Denitrification potential, Leach-out potential	

Fig. 15.4. Nitrogen-balance elements in RAUMIS.

$Input_{cf}$ = input of chemical fertilizer;
N_{req} = aggregated, regionalized requirement for N; and
N_{man} = manure nitrogen.

The input coefficients of chemical fertilizer are scaled so that the aggregated input (in relation to region and production activity) is equal to the observed fertilizer input in the base year. The regionalized N requirement is calculated in the intensity module (see above).

The input of N in manure varies with the kind of animal and the production level of each activity (dairy production, pork, etc.). In the case of dairy cattle, N output is a function of milk production. Losses of N, as ammonia from housed stock and during storage, constitute the 'ammonia losses' element in the N balance. Symbiotic fixation depends on the regional production levels of legumes, with the coefficient for beans, for example, calculated at 120 kg N ha^{-1}. N use by plants is determined from the plant's N content, information being derived from a variety of data sources. The N balance is derived by subtracting N output from N input.

Biodiversity and land-use diversity indicator

The implementation of an indicator of anthropogenic intervention in nature and landscape related to species and biotope protection, the RAUMIS Biodiversity Indicator (RBI), uses as a methodological basis the Hemeroby concept, developed by Blume and Sukopp (1976). This allows crop production to be ranked according to its importance for biodiversity under different levels of production intensity. The purpose of this indicator is to give a rough outline of the influence of agricultural land use on the environment, and to estimate the impacts of different policy measures: it should not be used for a detailed analysis of the agricultural and environmental situation as performed by land observation.

The RBI was developed following dialogue with 30 expert ecologists, the modelling group and geographers from the University of Bonn. Five separate citeria were defined for agricultural land use:

- I1: proximity of real and potential natural vegetation;
- I2: comparison of the 'real' conditions of agricultural land with its 'optimal' state in terms of biodiversity;

- I3: evaluation of endangered species and biotopes;
- I4: evaluation of arable land with respect to its function as biotope and biotope-connection;
- I5: time for regeneration for symbiosis after essential man-made influence.

The standard for comparison of the five elements is the typical state of the land cultivated by man in the first half of this century. The experts valued the RAUMIS crop activities, which were divided into four different levels of intensity, ranked between 'criterion not fulfilled' (0 points) and 'criterion totally fulfilled' (4 points). The levels of intensity were defined as follows:

- *extensive*: no chemical crop protection (CCP), low to average N fertilizer use;
- *less intensive*: low CCP, low to average N fertilizer use;
- *intensive*: average to high CCP, high N fertilizer use;
- *highly intensive*: high to very high CCP, high to very high N fertilizer use.

The weight of each criterion to a total indicator of biodiversity was calculated using the approach of constant sums (Torgerson, 1967). The values of each single indicator (1–5) are aggregated to one activity-specific total indicator 'importance for species and biotope protection'.

The weights for each single indicator as they were given by the experts are shown in Table 15.1. The result is an activity-based index for each RAUMIS region, detailed analyses of which can be found in Weingarten (1996). In the meantime, sensitivity analyses tested the reaction of the RBI for a set of newly defined production activities which represent less intensive production (important for reactions under more liberalized market and policy conditions).

The RBI approach differs from more precise, regionally disaggregated ecological analyses. Answering agricultural and environmental policy questions with a sector model forces a change in focus from the 'micro' to the 'macro' level. This approach can therefore be seen as 'first step' information in the decision-making policy process and a final step in the policy consulting process.

A refinement of the environmental module of RAUMIS was the implementation of the 'Shannon index' (Bork *et al.*, 1995). In this approach, the diversity of land-use systems in each region is examined. The definition of each main land-use system is essential for interpretation of later results. In the case described here, all the main crop activities of RAUMIS, including the set-aside

Table 15.1. Weights of the five RBI elements.

I1	0.17
I2	0.18
I3	0.25
I4	0.23
I5	0.17

programme and fallow land, are examined. Forestry, open water and land not in agricultural use were omitted.

The diversity of agricultural land use examined consists of two elements:

- number of land-use activities practised; and
- steady distribution of those activities in one region.

The resulting diversity of agricultural land use in a region is higher where more activities exist and the more steady their distribution is. To take into account both elements, the diversity index (*Hs*) according to Shannon was used.

$$Hs = -1 \times \sum_{i=1}^{n} \left(\frac{b}{a_i} \right) \times \ln \frac{b}{a_i}$$ [15.1]

where:

a is the total amount of land-use activities put into practice;
b is the realized level of each land-use activity; and
n is the total number of possible land-use activities of RAUMIS.

The size of the diversity index rises with the number of activities and their homogeneous distribution in one region. In our case the possible 31 main crop activities of RAUMIS lead to a maximum on the *Hs* index of 3.43. This would be the case, if exactly 3.23% of the arable land of one region is used by one of the 31 RAUMIS main land-use activities.

The relation of actual diversity (*Hs*) to maximal possible diversity (*Hs*$_{max}$) is called equidity or evenness (*E*) (Schäfer and Tischler, 1983).

$$E = \frac{Hs}{Hs_{max}}$$ [15.2]

The newly implemented Shannon index should be seen only as an addition to the already implemented index of biodiversity related to the concept of Blume and Sukopp (1976). It cannot provide information on the intensity of a single land-use activity. It is, without doubt, possible that a plot of wheat may be less harmful for the environment than a very intensively used meadow. Nevertheless, this explorative implementation of the diversity index shows the possibility of a fruitful further development of the environmental module of RAUMIS.

Greenhouse gas emissions module

The greenhouse gas inventory is still under construction. It should enable the influence of different policy measures on greenhouse gas emissions in German agriculture to be shown. The gases studied are: methane (CH_4), ammonia (NH_3), carbon dioxide (CO_2) and nitrous oxide (N_2O). It is planned that the contribution of agriculture to global warming should be calculated and the effects of different environmental policy measures such as energy tax or tradable emission rights on agricultural income as well as production structure and input–output relations demonstrated.

At the moment, there are coefficients for each animal activity implemented. To determine an approximate level of emissions, the coefficients are multiplied by each level of animal production. So far, there are equal sectoral coefficients, which do not take into account different systems and techniques of application or storage of manure. Technical progress in herbage conservation techniques is also not considered.

APPLICATION

This section contains a brief analysis of the impacts of a liberalized CAP on the environment. The target year for the analysis is 2005. The analysis is a comparative static comparison of a reference run and a simulation run. After a short scenario specification, the effects of decoupling CAP subsidies on quantities and production structure are presented. The presentation of model results is done in such a way that results of the reference run are compared with observed quantities and the production structure of 1995/96; model results of the decoupling scenario (simulation run) are compared to results of the reference run. The section is followed by a regionalized impact analysis of selected environmental variables of RAUMIS.

Scenario Specification

The scenario was specified with experts of the Federal Ministry of Food, Agriculture and Forestry. The vector of prices stems from a dynamic coupling process with a non-spatial partial equilibrium market model developed by the Federal Agricultural Research Centre and described by Frenz *et al.* (1995).

The main political context is that world market prices for cereals will rise to the EU level until the target year 2005. The assumption is that the EU will be able to export cereals on world market without export subsidies.

The *reference run* has as its political frame actual CAP political measures. CAP subsidies in the target year for crop and animal production are nominally constant with those of 1996. The minimum percentage for the set-aside programme is 5%. It is possible to set farmland aside to a maximum of 33% of arable land. The restriction of 5% for minimum set-aside results from a careful interpretation of the GATT (General Agreement on Tariffs and Trade) Treaty by the Federal Ministry concerning CAP subsidies conformity with the GATT. It is assumed that quantity-regulating measures have to be used so that the blue-box character of CAP subsidies is guaranteed.

The *simulation run* is equal to the reference run with respect to prices, quotas, etc. The differences are that all CAP subsidies are paid decoupled and that no set-aside programme exists.

Model Results

When an 'optimistic' price level is assumed, oilseed production stagnates compared with 1996, at a level of 785 000 ha in the *reference run*. This is close to the net-guaranteed cropping area of the Blair House agreement for oilseeds (about 836 000 ha reduced by 10% from 929 000 ha). As a result of an annual increase of 1.4% in yields and an increase in cereal cropping area of about 20%, German production of cereals in the reference run is supposed to rise to 54 Mt (from about 39 Mt in 1996). The Federal Ministry estimates a tremendous reduction in German beef production. That leads to a scenario specification for the reference run in which German beef production will be reduced by about 16%, down to 1.18 Mt (from about 1.4 Mt in 1996). The production of pork stagnates, while that of poultry meat rises to 665 000 t (+10% compared with production in 1996).

The *decoupling* of CAP subsidies reduces the distortion of competitiveness of agricultural production activities. In cereal production, decoupling leads primarily to an increase of soft wheat cropping. It rises from 2.96 Mha in the reference run to 3.05 Mha in the decoupling scenario. The oilseed cropping area is reduced from 785 000 ha to 222 000 ha. On the other hand, fallow land rises dramatically to 1.46 Mha (from about 319 000 ha in the reference run), the highest increase of fallow land being found in east Germany (902 000 ha).

The magnitude of the changes results primarily from a change in production structure, but also from increases in average yield. The abolition of the obligatory set-aside programme leads to a shift of production from marginal areas to more favourable areas. This explains why increasing yields and increasing acreage of cereal cropping are leading to a further increase in German cereal production to 55.2 Mt (+2.3%). The abolition of oilseed subsidies causes a tremendous reduction of oilseed production down to 811 300 t (−71%) and pulses (−10% to 373 000 t). The main effect of decoupling on animal production is a further 2% reduction of beef production, while effects on pork and poultry meat production are marginal. Decoupling of subsidies leads to a reduction of direct CAP subsidies (−81% compared with 8.9 billion Deutschmarks (DM) in the reference run) and a decrease of net value added (−24% compared with 27.5 billion DM in the reference run). The biggest loser concerning payment of direct CAP subsidies is Sachsen-Anhalt. Here, a decoupling of CAP subsidies leads to the result that only 11% of the subsidies of the reference run are paid to farmers in the simulation run. The payments after decoupling CAP subsidies are, for example, 'gas oil reduction payment' or 'less favoured area payment' (national subsidies). Rheinland-Pfalz has the lowest reduction in subsidies for agriculture. Here, 26% of the reference run payments are still coming to the farms (high proportion of disadvantaged area).

The *environmental indicators* in RAUMIS depend on the endogenously calculated intensity and production structure. Therefore, the environmental effects from agricultural production can be measured directly.

In the reference run, the N balance for Germany shows a surplus of about 75 kg N ha^{-1} farmland. It varies from 85 kg in the west to 55 kg in the east. Compared with information from the base-year run for 1991, there is little difference (74 kg, 85 kg and 49 kg) for the following three reasons:

- Firstly, the reduction in *animal production* from 1991 to the target year 2005 leads to a reduction in N from manure of about 17 kg ha^{-1} farmland (from 86 kg in 1991).
- As a result of this, the input of *mineral fertilizer* rises from 110 to 121 kg ha^{-1} farmland to fulfil N requirements.
- The third effect that influences N balance results from a change in production structure. In the target year, the share of *extensive production* activities, especially less intensive production on grassland, rises markedly. That leads in the reference run to an average reduction in direct requirement of N by plants of 4 kg, to 129 kg ha^{-1} German farmland.

The regional distribution of N surplus in the reference run is shown in Fig. 15.5. Even after a reduction in animal production, the typical regions for animal production in the west of Germany on the border with The Netherlands can be identified very easily by their high N surplus that results from manure N. Also, the typical grassland area of Nordrhein-Westfalen (Süderbergland), with a N surplus below 50 kg ha^{-1} farmland, is notable; the eastern part of Germany has generally lower N surpluses than the western part. In particular, the southern parts of east Germany (Thüringen, Sachsen) have an average nitrogen surplus below 50 kg, due to a relatively high share of grassland and reduced animal production.

Figure 15.6 shows the average nitrogen surplus per hectare of farmland on NUTS 1 level of Germany (Bundesländer) excluding Hamburg, Bremen and Berlin. All NUTS 1 regions, except Nordrhein-Westfalen, show a decrease in N surplus in the decoupling scenario. The ranking between regions in the two target year runs is mainly the same. Nordrhein-Westfalen has the biggest N surplus, because of intensive animal production in its north-western parts. The input of nitrogen from manure per hectare of farmland in Nordrhein-Westfalen stagnates at about 100 kg in both target-year runs.

Most remarkable is the reduction of N surplus in the regions of the former German Democratic Republic, for three reasons: first, the reduction in animal production; secondly, the rise of the share of less intensive crop production (shown in Fig. 15.7); and thirdly, the tremendous rise in fallow land. Sachsen, the area with the smallest N surplus, is characteristic of the eastern regions. Nearly a quarter of its arable land in the decoupling scenario is cultivated with less intensive crops. Elsewhere, the share of fallow land on farms rises from 1.7% in the reference run to nearly 15% (about 12 000 ha) in the simulation run.

In both the reference run and the simulation run, the share of less intensive crop production activities in commercial crop production in eastern Germany is much higher than in the western part (see Fig. 15.7). This is caused

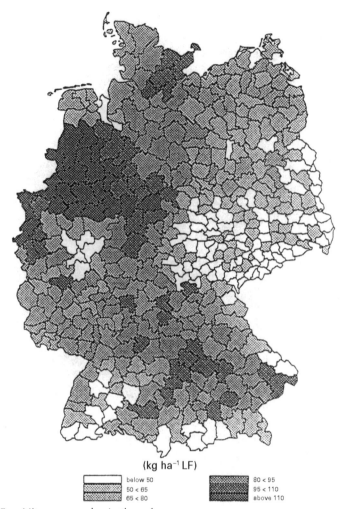

(kg ha⁻¹ LF)

	below 50		80 < 95
	50 < 65		95 < 110
	65 < 80		above 110

Fig. 15.5.   Nitrogen surplus in the reference run.

primarily by the eastern region's typically high share of 'expensive' non-familial working units and partly by a lower soil classification (e.g. Brandenburg) than in Germany's western regions. The lower N surplus (Fig. 15.5) and lower-intensity crop production lead to higher values on the RAUMIS biodiversity indicator. Concerning the diversity of production structure (expressed in the Shannon index), the tremendous rise in fallow land caused by a decoupling of subsidies leads to a loss of diversity in all NUTS 1 regions of Germany, with emphasis on Germany's eastern regions. Mecklenburg-Vorpommern reports the highest value of evenness (shares of main crop activities on farmland in that region are nearly equal, see equation 15.2) in the

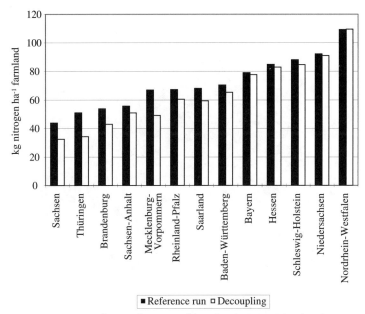

Fig. 15.6. Nitrogen surplus per hectare of farmland on NUTS 1 level.

reference run (75%). This value is reduced to 70% in the simulation run, mainly caused by an increase in fallow land.

CONCLUSIONS

The main objective of this chapter has been to give an outline of the methodological aspects of RAUMIS, illustrated by a brief example of how the model can be used for detailed impact analyses of special policy measures. It is of note that decoupling CAP subsidies as specified in the simulation run does not have substantial effects on the analysed environmental variables of RAUMIS. That may result from objectives of policy-makers, who mainly specified the scenario for a more detailed economic analysis of decoupling.

In general, the RAUMIS modelling system is an example of how mainly economics orientated models can be complemented by additional information on several ecological variables, such as balances or potential importance for biodiversity. The development of RAUMIS from its inception in the late 1980s was always orientated towards customer demand, with the aim of creating a practical and flexible tool for policy consultation.

The deep regional differentiation of an agricultural sector model such as RAUMIS was necessary for the depiction of the increasing regionalization of agricultural and environmental policy measures (quotas, set-aside

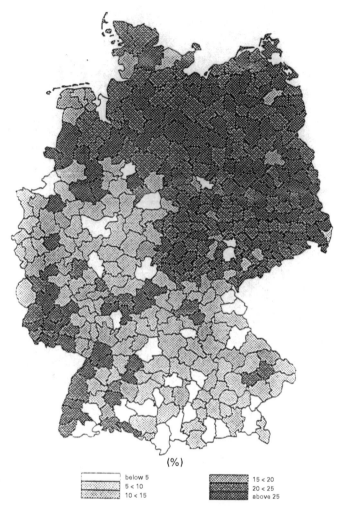

(%)

	below 5		15 < 20
	5 < 10		20 < 25
	10 < 15		above 25

Fig. 15.7. Extensive crop production in the simulation run.

regulations, direct income transfers, etc.). Further regionalization would probably be doomed to failure because of the enormous data need.

On the whole, the possibilities inherent in regionalization and in incorporating environmental indicators confirm that the modular approach of RAUMIS allows flexible (and technically straightforward) adjustments to be made to changing problems and issues relevant to policy-making.

Use of the model for the Federal Ministry of Food, Agriculture and Forestry and other organizations has demonstrated that the principal idea of staying close to actual demand has contributed to the success of RAUMIS both now and in the future.

REFERENCES

Bauer, S. and Kasnakoglu, H. (1990) Non-linear programming models for sector and policy analysis. *Economic Modelling* 1, 275–290.

Blume, H.P. and Sukopp, H. (1976) Ökologische Bedeutung anthropogener Bodenveränderungen. *Schriftenreihe für Vegetationskunde* 10, 83–85.

Bork, H.-R., Dalchow, C., Kächele, H., Piorr, H.-P. and Wenkel, K.-O. (1995) *Agrarlandschaftswandel in Nord-Ostdeutschland unter veränderten Rahmenbedingungen: Ökologische und ökonomische Konsequenzen, Berlin*. Ernst & Sohn, Berlin, pp. 286–312.

Cypris, C. (1998) Positive Mathematische Programmierung im Agrarsektormodell RAUMIS. Dissertation forthcoming, Schriftenreihe der Forschungsgesellschaft für Agrarpolitik und Agrarsoziologie e.V., Bonn.

Frenz, K., Manegold, D. and Uhlmann, F. (1995) EU-Märkte für Getreide und Ölsaaten. Künftige Entwicklungen in der Erzeugung und Verwendung von Getreide, Hülsenfrüchten und Ölsaaten in der Europäischen Union. *Schriftenreihe des Bundesministeriums für Ernährung, Landwirtschaft und Forsten, Reihe A: Angewandte Wissenschaft*, H. 439.

Howitt, R.E. (1995) Positive mathematical programming. *American Journal of Agricultural Economics* 77, 329–342.

Löhe, W. (1998) *Extensivierungspotentiale in der Landwirtschaft*. Shaker, Aachen.

Schäfer, M. and Tischler, W. (1983) *Wörterbuch der Biologie – Ökologie*. Fischer, Stuttgart.

Torgerson, W. (1967) *Theory and Methods of Scaling*. Springer, New York.

Weingarten, P. (1996) *Grundwasserschutz und Landwirtschaft. Eine quantitative Analyse von Vorsorgestrategien zum Schutz des Grundwassers vor Nitrateinträgen*. Vauck, Kiel.

Criteria and Indicators: Experience in the Forestry Sector

16

Ewald Rametsteiner

INTRODUCTION

Forests are among the largest, most diverse and complex ecosystems in the world, covering about 27% of the world's land surface. In Europe (including eastern Europe but excluding the former USSR), about 30% of the total land area (470 million ha) consists of forests and woodland (FAO, 1997). They range from sub-Mediterranean cork oak forests in the south to boreal birch forests in the north and from high alpine mountain to coastal forests.

The majority of the world's species are thought to be located in forests. Species diversity is much greater in tropical forests than in temperate forests. However, the genetic diversity within species in temperate forests is estimated to equal that in tropical forests (Dudley, 1992).

Forest products (both wood and non-wood) are an important resource as an input for a great variety of economic activities. These range from the production of fuelwood and roundwood for industrial purposes to wood for musical instruments, and from deer and other animals to fruits, mushrooms and herbs. Furthermore, forests fulfil other essential functions for society, including protection from adverse natural impacts such as landslides or floods; they also play an essential role in water and atmospheric cycles and provide recreational opportunities.

Many of these ecological, economic and social characteristics have been measured and assessed in a variety of contexts, including policy and planning at various levels, evaluation and research. In recent years, however, an unparalleled international interest has emerged in characterizing the interest of society in forests and/or their functions and establishing indicators (and related criteria) for their measurement.

©CAB INTERNATIONAL 1999. *Environmental Indicators and Agricultural Policy* (eds F.M. Brouwer and J.R. Crabtree)

The Global Background

The last two decades of the twentieth century saw a rising threat to the quality and quantity of forests on a global scale. In the tropics the rate of deforestation was drastically increasing, resulting in concern about matters such as the high losses of biological diversity, increasing amounts of CO_2 release into the atmosphere and the fate of indigenous forest dwellers. In Europe the forests faced considerable threats of declining health and vitality. Consequently, public and political interest in, and concern about the situation of, forests increased throughout the world.

The political community was challenged to elaborate solutions to meet the diverse demands of society with regard to forests and, at the same time, to preserve the integrity of forests in qualitative and quantitative terms. International high-level conferences on forest policy directed towards the elaboration of solutions and the improvement of cooperation and concerted action amongst countries have been held in increasing numbers since the beginning of the 1990s. The most important international conference was the UN Conference on Environment and Development (UNCED), held in Rio de Janeiro in 1992. At UNCED two of the major outcomes were mainly related to forests and their proper management:

- The 'Non-Legally Binding Authoritative Statement of Principles for a Global Consensus on the Management, Conservation and Sustainable Development of All Types of Forests' ('Forest Principles'). These principles, although not legally binding, comprise the first global consensus about principles of forest management. The political goal that was adopted by these principles was 'sustainable forest management', already well known and applied in some parts of the world.
- 'Agenda 21, Chapter 11: Combating Deforestation'. The governments agreed to pursue in cooperation with interest groups and other international organizations 'the formulation of scientifically sound criteria and guidelines for the management, conservation and sustainable development of all types of forests' (Agenda 21, Chapter 11, Section 11.22 (b)).

The follow-up process of the UNCED conference and especially of the implementation of Agenda 21 is administered by the UN Commission on Sustainable Development (UN-CSD). In the follow-up it became clear that the international policy and political debate on forests should be given a broader political forum that focused on priority issues. The CSD established an *ad hoc* Intergovernmental Panel on Forests (IPF) in 1995, one of whose five priority issues was 'Scientific research, forest assessment and development of criteria and indicators for sustainable forest management (SFM)'.

The European Background

In Europe, two main factors were the driving forces behind recent progress in the development of, and/or use of, indicators in forestry: first, the increasing threat to the health and vitality of European forests that led to the establishment of a pan-European process on forest policy by forest ministers in 1990, and secondly, the need for a follow-up activity for the UNCED process. The most important step towards setting common goals for SFM in European forests took place at the Second Ministerial Conference on the Protection of Forests in Europe in July 1993. At this ministerial conference, which was attended by 34 European countries and the European Community, *inter alia*, two resolutions were adopted:

- Resolution H1: General Guidelines for the Sustainable Management of Forests in Europe.
- Resolution H2: General Guidelines for the Conservation of Biodiversity of European Forests.

In the course of this intensive political process to elaborate these resolutions a new understanding and definition of SFM was acknowledged, where the meaning of the term was expanded from a primarily economic concept to one with a broader perspective embracing ecological and social dimensions. Indicators were identified as appropriate tools for further concretization and clarification of the concept of SFM for monitoring sustainability in forest management, and for the evaluation of measures for its improvement.

INDICATORS IN FORESTRY – THE THEORETICAL FRAMEWORK

Recent efforts to elaborate indicators in the field of forestry are almost exclusively connected with the measurement and reporting of the degree of the sustainability of forest management. The starting point for the elaboration of a suitable measurement framework is a definition of what SFM is about.

Defining Sustainable Forest Management

SFM is a complex normative concept which today embraces three main dimensions: ecological, economic and social. With regard to defining SFM, several partial interests or objectives have to be taken into account in order to determine what exactly should be utilized and/or preserved through time (Fig. 16.1).

A definition of SFM is thus the result of a political process driven by the various actors, their values, interests and their negotiating power. The definition adopted by the European forest ministers in 1993 in Preamble D of Resolution H1 reads:

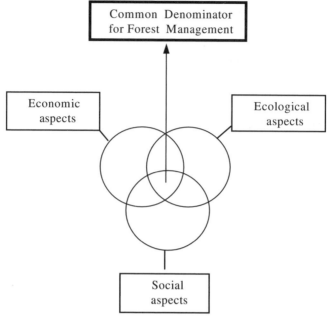

Fig. 16.1.   Diverging interests and values to be considered in the management of forests.

> sustainable forest management means the stewardship and use of forests and forest lands in a way and at a rate, that maintains their biodiversity, productivity, regeneration capacity, vitality and their potential to fulfil, now and in the future, relevant ecological, economic and social functions, at local, national, and global levels, and that does not cause damage to other ecosystems.
>
> (Ministry of Agriculture and Forestry, 1995).

The Operationalization of Sustainable Forest Management

The general approach
In order to further operationalize the rather abstract definition of SFM, a hierarchical system of principles/guidelines, criteria and indicators has generally been adopted in forestry. Although these elements or instruments are not applied consistently by the various actors within the sector, they can be found in most of the initiatives (Maini, 1993; Prabhu *et al.*, 1996; Bueren and Blom, 1997). Figure 16.2 shows the instruments as theoretically applied in rational planning within a policy cycle.

   Principles are used to lay down general objectives at an abstract level. A widely used definition of a principle is 'a fundamental truth or law as the basis

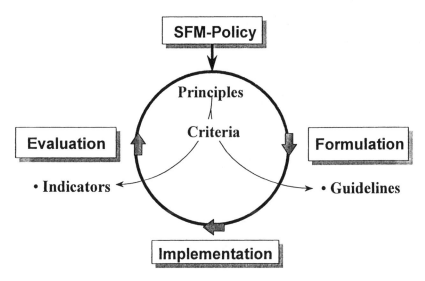

Fig. 16.2. The instruments of rational planning within the policy cycle.

of reasoning or action'. Guidelines are used to formulate actions that should ensure the achievement of the overall goal of SFM.

Criteria are commonly defined as 'distinguishing aspects that are considered important and by which a subject can be assessed'. Criteria are used as a tool for evaluating the attainment of a specific objective related to SFM. At the same time, they are a tool for further defining and clarifying what is meant by SFM. A common definition and consistent use of the term 'criterion' is lacking in forestry, which creates considerable problems in political negotiation processes. Two types of approaches can frequently be found: first, the criterion as a standard/objective-setting tool where a criterion defines something that has to be attained, e.g. the maintenance of forest health; and secondly, the criterion as an aspect in relation to which measurement takes place, e.g. quantity of water resources.

Indicators generally act as signs or symptoms for the presence of something. They show something by making use of data. The measurement of (data for) indicators allows the assessment of the state of the art, and continuous monitoring allows the detection of change. As discussed above, recent efforts to elaborate indicators in the field of forestry are almost exclusively connected with the measurement of the sustainability of forests and/or forest management.

Indicators link the theoretical or conceptual level of SFM with 'reality'. Indicators are thus connected to data measured 'in the field'. In those cases where a single result or a few condensed evaluation results (multicriteria evaluation) are the desired output, a system of aggregation and weighting is applied to derive indices.

THE SELECTION OF INDICATORS.   Indicators are applied on a variety of scales. Some, like Ott (1978), define indicators on an aggregate scale and use the term identically with 'index'; others link an indicator to one specific type of data (see also Mitchell, 1996). According to their aggregation level, indicators can be divided into three groups:

- single-data indicators;
- key indicators of one or more type of data;
- single-index indicators.

Which aggregation level to choose depends on the target group of the indicator:

- indicators for the public necessitate a high aggregation level;
- indicators for policy makers require medium aggregation level;
- indicators for scientists necessitate a low aggregation level.

In theory, indicators to be selected have to fulfil the basic requirements of reliability, validity and effectiveness of resource use (time, money). However, ideal indicators are difficult to locate and several constraining factors typically play a role in their selection.

Indicators for Measuring the Attainment of Sustainable Forest Management

Based on the definition and structure of SFM, adequate indicators are required for the ecological, economic and social dimensions, if SFM is to be measured adequately.

INDICATORS FOR THE ECOLOGICAL DIMENSION.   Forest   ecosystems   typically show a high degree of diversity and a high complexity of dynamic functional interactions. The level of knowledge on forest ecosystems is, however, far from complete. The comprehensive measurement of an object of interest such as an ecosystem, its functioning and reaction to influences, is extremely demanding and the principal method that is applied to arrive at practical solutions is one of complexity reduction. However, in order to get useful results it is of crucial importance to reduce complexity without loss of validity or reliability. The approach often taken is to identify key ecosystem elements and processes.

A further step in complexity reduction consists of pattern recognition, i.e. patterns on a temporal and spatial scale, and patterns of forest composition (forest types). Noss (1990) distinguished three features within each level of organization (i.e. genetic; species; community, ecosystem and landscape): first, composition; second, structure; and third, function. This approach, which differs from that followed previously, in which key indicator species were identified as representative indicators for whole habitats, has been discussed extensively in recent years. It has to be said, however, that considerable research has still to be undertaken, and existing research collected and integrated, in order to devise reliable measurement and monitoring systems for forest ecosystems.

The ultimate goal of indicators for SFM is the determination of threshold levels for human impacts on, or human utilization of, forests. The measurement of human impacts on forest ecosystems, however, faces the same difficulties as the measurement of the functioning of ecosystems in general: knowledge is limited, and little is known, for example, on the preconditions for the stability and resilience of forest ecosystems that are necessary if threshold levels for allowable impacts are to be determined (see also Rees, 1994; OECD, 1995).

Again, complexity reduction is the principal approach taken. Despite the diverse structure of the elements considered for measurement of SFM, impacts are generally addressed using a single impact matrix, such as in Table 16.1. The indicator matrix is adapted to the application to indicate only those impacts that can actually be influenced by the actors in question. Forest managers, for instance, are not able to influence effectively impact factors such as air pollution.

INDICATORS FOR ECONOMIC AND SOCIAL ASPECTS. The measurement of direct economic aspects in relation to SFM (e.g. economic output) is comparatively straightforward, since the elements and interactions are easy to observe and measurable (see also Saleth, 1994). Although more complex, the same is true for the measurement of direct and indirect effects of forestry on society, for example, the effects of forest management activity on recreation or on the protective role of forests. Indicators devised for economic or social aspects are often located at their interface ecological dimension. The indicator for maximum allowable output (annual allowable cut, AAC) of wood, for instance, is located between the economic and the ecological systems, as the AAC is mainly determined by the net primary production of woods. (The use of indicators for maximum output of non-wood products, however, is again hampered by the

Table 16.1. Matrix of indicators for the measurement of ecological impact aspects on forest ecosystems.

	Impact from		
Impact on	Forestry	Society	Other
Ecosystem elements			
soil			
water			
biodiversity			
genetic			
species			
ecosystem			
Ecosystem processes			
'health and vitality'			

lack of ecological knowledge on growth and regeneration patterns of many of these products.)

Approaches differ, but many follow a utilitarian concept of 'functions' provided by forests to society. These functions, categorized into productive, protective and other (e.g. recreation), are measured by indicators in terms of the quantity and quality of the functions or output provided (Table 16.2).

Common International Indicator Sets

The application of a common set of indicators in an international context creates some specific problems that have to be considered. In Europe, all countries have established information and monitoring systems for forestry. However, these systems, and the data derived, differ in the use of terminology and the types of variables measured as well as in the methods adopted – including differences, for example, in sample design, time scales, statistical precision and plot system. Common evaluation and reporting systems thus face both political and technical difficulties in adjusting and harmonizing the individual national or sub-national systems and approaches (Ministry of Agriculture and Forestry, 1996a, b).

INDICATORS IN FORESTRY: PRACTICAL APPLICATIONS

Two main applications of indicators currently dominate in forestry. Both are related to the achievement of SFM:

- a top-down approach: national and international forest policy at national government level;

Table 16.2. Matrix of indicators for the measurement of output aspects of sustainable forest management

Output factor	Actual output
Productive functions	
wood products	
non-wood products	
Protective functions	
soil erosion	
water protection	
Other functions	
recreation	
etc.	

- a bottom-up approach: certification of sustainably managed forests at sub-national or forest management unit level.

Indicators in Forest Policy

Introduction
The approach of using criteria and indicators for policy formulation, as well as for evaluation and reporting of achievements, emerged soon after the UNCED. A Seminar of Experts on the Sustainable Development of Temperate and Boreal Forests was held in Montreal in September 1993 under the aegis of the Conference of Security and Co-operation in Europe (CSCE). This high-level scientific and technical forum examined the scientific basis for the conservation, management and development of temperate and boreal forests. The major outcome of the seminar was a preliminary set of criteria and some potential indicators for SFM elaborated primarily on a scientific basis (Maini, 1993). This set served as an indicative guide and as background material for several other initiatives.

The elaboration of criteria and indicators within the Helsinki Process
In Europe, the follow-up process for the Resolutions H1 (General Guidelines for the Sustainable Management of Forests in Europe) and H2 (General Guidelines for the Conservation of Biodiversity of European Forests) of the Second Ministerial Conference on the Protection of Forests in Europe in July 1993 (the Helsinki Process) also started to consider criteria and indicators (C&I) for evaluation and reporting in 1993. Subsequently, a set of C&I was elaborated by the administrative staff of the secretariat of the process, the so-called Liaison Unit.

The first important milestone of the political follow-up process to the resolutions was achieved in 1994 with the adoption of six criteria and 27 most suitable quantitative indicators by the national representatives of the signatory states. The six criteria, adopted after a political negotiation process, represent the political consensus achieved by European countries on the most important characteristics, at the conceptual level, of SFM. The relevant quantitative indicators are measurable parameters for evaluating the fulfilment of the six criteria at a national level. These criteria and indicators cover social, cultural, economic and ecological elements. Table 16.3 presents an overview of the criteria and indicators adopted by the Helsinki Process (Ministry of Agriculture and Forestry, 1994).

In order to follow up the C&I, and investigate the applicability of the quantitative indicators, a questionnaire was designed to test measures of the 27 adopted quantitative indicators of SFM. The results of this questionnaire, which was sent out to 39 countries in September 1994, represented the first attempt to describe the present status of SFM in Europe. Responses to the criteria differed between countries and zero responses to some criteria by some

Table 16.3. Quantitative criteria and indicators used in the Helsinki Process to assess sustainable forest management on a national scale.

Criteria	Indicators
1. Maintenance and appropriate enhancement of forest resources and their contribution to global carbon cycles	1.1 Area of forest and other wooded land and changes in area (classified, if appropriate, according to forest and vegetation type, ownership structure, age structure, origin of forest) 1.2 Changes in: (a) total volume of the growing stock (b) mean volume of the growing stock on forest land (classified, if appropriate, according to different vegetation zones or site classes) (c) age structure or appropriate diameter distribution classes 1.3 Total carbon storage and changes in the storage in forest stands
2. Maintenance of forest ecosystem health and vitality	2.1 Total amount of and changes over the past 5 years in depositions of air pollutants (assessed in permanent plots) 2.2 Changes in serious defoliation of forests using the UN/ECE and EU defoliation classification (classes 2, 3, and 4) over the past 5 years 2.3 Serious damage caused by biotic or abiotic agents: (a) severe damage caused by insects and diseases with a measurement of seriousness of the damage as a function of (mortality or) loss of growth (b) annual area of burnt forest and other wooded land (c) annual area affected by storm damage and volume harvested from these areas (d) proportion of regeneration area seriously damaged by game and other animals or by grazing 2.4 Changes in nutrient balance and acidity over the past 10 years (pH and CEC); level of saturation of CEC on the plots of the European network or an equivalent national network
3. Maintenance and encouragement of productive functions of forests (wood and non-wood)	3.1 Balance between growth and removal of wood over the past 10 years 3.2 Percentage of forest area managed according to a management plan or management guidelines 3.3 Total amount of and changes in the value and/or quantity of non-wood forest products (e.g. hunting and game, cork, berries, mushrooms, etc.)

Table 16.3. *contd.*

Criteria	Indicators

4. Maintenance, conservation and appropriate enhancement of biological diversity in forest ecosystems	4.1 Changes in the area of:
	    (a) natural and ancient semi-natural forest types
	    (b) strictly protected forest reserves
	    (c) forests protected by special management regime
	4.2 Changes in the number and percentage of threatened species in relation to total number of forest species (using reference lists, e.g. IUCN, Council of Europe or the EU Habitat Directive)
	4.3 Changes in the proportion of stands managed for the conservation and utilization of forest genetic resources (gene reserve forests, seed collection stands, etc.); differentiation between indigenous and introduced species
	4.4 Changes in the proportion of mixed stands of 2–3 tree species
	4.5 In relation to total area regenerated, proportions of annual area of natural regeneration

5. Maintenance and appropriate enhancement of protective functions in forest management (notably soil and water)	5.1 Proportion of forest area managed primarily for soil protection 5.2 Proportion of forest area managed primarily for water protection
6. Maintenance of other socio-economic functions and conditions	6.1 Share of the forest sector from the gross national product 6.2 Provision of recreation: area of forest with access per inhabitant, % of forest area 6.3 Changes in the rate of employment in forestry, notably in rural areas (persons employed in forestry, logging, forest industry)

countries were mainly explained as follows (Ministry of Agriculture and Forestry, 1995):

- complete lack of data;
- difficulties in obtaining data in the time scale required;
- inadequate definition of questions and/or guidelines.

As some aspects were felt to be inadequately covered by the quantitative indicators adopted in 1994, the pan-European Process additionally adopted the following descriptive indicators as a supplement in 1995:

- legal/regulatory frameworks;
- institutional frameworks;

- financial instruments;
- informational means.

The descriptive indicators serve as a way of assessing those aspects of SFM that can be appraised only through the existence of a policy instrument and its effective implementation.

As a whole, the C&I adopted serve not only as an evaluation and information tool, but also as a guide to the implementation of national policies and action plans for the sustainable management of forests and the conservation of their biodiversity.

The objectives that determined the selection of the C&I differed considerably between signatory states in Europe. From the final list of adopted indicators it can be seen that a common principle was to use existing data for reporting to the public rather than to invest in new resource monitoring systems. No rearrangement of information systems and no assessment of new data was specifically asked for in the Helsinki Process, nor is the provision of information for these C&I binding. The primary function of the Helsinki Process C&I therefore focuses on information diffusion to the public through concerted voluntary reporting. The C&I of the Helsinki Process also reinforce and harmonize the understanding of what constitutes SFM within European countries at a political level.

The Montreal Process and other initiatives

A similar process to the Helsinki Process was started by governments of the temperate and boreal belt outside Europe in 1994. The Working Group on Criteria and Indicators for the Conservation and Sustainable Management of Temperate and Boreal Forests (the Montreal Process) was formed to advance the development of internationally agreed C&I for the conservation and sustainable management of temperate and boreal forests in other regions outside Europe at the national level. In February 1995 the countries represented in this group endorsed a comprehensive set of C&I for forest conservation and sustainable management for use by their respective policy-makers (Montreal Process, 1996).

One of the main differences between the Helsinki Process and the Montreal Process concerns the level of political commitment to use the C&I in the respective policies of the participating countries. Although the reporting on C&I is not legally binding in the Helsinki Process, the degree of political commitment is considerably higher, mainly because of the follow-up character of the C&I to the binding resolutions that were signed by the forest ministers in Helsinki in 1993. The Montreal Process is not based on a similarly strong political commitment. As a result, the Helsinki Process is characterized by a higher political commitment and the Montreal Process by a higher technical/scientific level of the discussion on indicators. A negative correlation can thus be observed between the level of political commitment and the quality and/or quantity of indicators in question. On the other hand, it will not be possible to implement the majority of indicators discussed within the Montreal Process because of resource restraints and a lack of political will in the majority of countries.

Several other regional initiatives were established in the follow-up phase in other regions of the world. Table 16.4 gives an overview of ongoing international initiatives on the elaboration and implementation of national-level criteria and indicators for SFM. It clearly indicates the widespread international interest in and political support for the development and implementation of SFM criteria and indicators.

Conclusion

The measurement of SFM by means of C&I has gained global political interest and support. It can be said that the international forest policy arena globally, and especially in Europe, is the most advanced sector in terms of using a concise and common system of C&I to measure key elements for sustainable development and to report the results to the public. As others are currently in the process of setting up similar common systems for measuring aspects related to sustainability, the experience gained in forestry is a highly valuable cross-sectoral input.

Criteria and Indicators for Forest Certification

A second broad field of application of C&I in forestry is the certification of sustainably managed forests, primarily at a forest management unit level.

Table 16.4. Coverage of ongoing initiatives on criteria and indicators for sustainable forest management by ecological region (source: UNCSD-IPF, 1996; author's own data).

Ecological region and initiative	Actors	Number of countries
Temperate and boreal forests		
Helsinki Process (including Russia[a])	Governmental	38
Montreal Process (including Russia)	Governmental	12
Tropical forests		
ITTO Producer Countries[b]	Governmental	25
Tarapoto Proposal	Governmental	8
Central America	FAO/UNEP	7
Dry-zone forests		
Sub-Saharan Dry-Zone Africa	FAO/UNEP	27
Near East	FAO/UNEP	20

[a] The number of countries for Helsinki refers to the signatory states of the Helsinki Resolution as well as those countries that have subsequently participated in the work of the Helsinki Process.
[b] International Tropical Timber Organisation.

Since the call by some environmental groups for a boycott of tropical timber to prevent rapid deforestation in the tropics was seen as undesirable or ineffective by others, efforts were undertaken to find alternative approaches. Positive incentives for the certification of sustainably managed forests and subsequent labelling of sustainably produced timber were explored. This market-based instrument was first brought into debate by non-profit organizations in the United States at the end of the 1980s, whereas in Europe the issue was first brought to public attention in Great Britain at the beginning of the 1990s. Since then, a multitude of certification initiatives has evolved worldwide, resulting in a proliferation of definitions and certification standards, including criteria and indicators for assessing SFM.

In 1993, a Forest Stewardship Council (FSC) was established in an attempt to harmonize certification schemes and establish a global framework of principles and criteria for SFM and to act as an accreditation body for certification organizations. The FSC has mainly an environmental and social NGO background. In 1995, the ISO was brought into the debate by the industry to act as a standard-setting body for specific sector standards on SFM within the ISO 14.000 series on environmental management systems. It subsequently started work on respective standards. Furthermore, some governments have been or are involved in the process of setting standards for certification of sustainably managed forests.

As the C&I applied in the context of certification are used for assessing aspects of sustainability of forest management on the forest management unit rather than on national levels, the indicators are specifically tailored to measure those aspects that are relevant at that level. However, due to the diversity of the schemes that are proposed and/or applied for assessment, no common set of indicators is currently in use.

Two types of 'standards' can generally be distinguished:

- process standards (and corresponding indicators) lay down specifications for management (organizational) processes which define how something should be done;
- performance standards (and corresponding indicators) lay down specific performance targets or norms that define the outcome that has to be achieved.

In general, the schemes that certify according to performance standards follow the same type of evaluation matrix as presented in Tables 16.1 and 16.2. As regards the impact factors, however, they concentrate on those factors that are relevant for the management of single forestry units.

The issue of forest certification and its potential for improving or encouraging SFM is still under debate. Nevertheless, considerable efforts are being made by diverse groups and these have resulted in a proliferation of certification standards as well as certification and assessment bodies. These developments have led to the non-standardized use of a broad range of indicators for assessing whether or not a forest is sustainably managed. As the market for inspection and assessment is determined by market factors such as competitive prices for assessment services, the risk of cost-derived reductions in indicator quality and quantity is high.

CONCLUDING REMARKS

The recent concentration of effort in the development of indicators in forestry (including those for the environment) has its roots in the political goal of SFM. Considerable progress has been made in the measurement of the relevant sustainability dimensions within the sector. However, technical (lack of knowledge), organizational (lack of resources) and political (diverging political interests) constraints have still to be overcome before the goal of measuring SFM can be achieved. Any conclusion on the capability of C&I to fulfil these demands would, at this stage, be premature.

REFERENCES

Bueren, L. and Blom, E. (1997) *Hierarchical Framework for the Formulation of Sustainable Forest Management Standards.* The Tropenbos Foundation, Backhuis, Leiden.

Dudley, N. (1992) *Forests in Trouble: A Review of the Status of Temperate Forests Worldwide.* WWF International, Gland.

FAO (1997) *State of the World's Forests 1997.* FAO, Rome.

Maini, J.S. (1993) A systematic approach to defining criteria, guidelines and indicators. Paper presented at the CSCE Seminar of experts on sustainable development of boreal and temperate forests. CSCE Seminar Paper, Montreal, Canada.

Ministry of Agriculture and Forestry (1994) *European Criteria and Most Suitable Quantitative Indicators for Sustainable Forest Management.* Ministry of Agriculture and Forestry Liaison Unit, Helsinki.

Ministry of Agriculture and Forestry (1995) *Interim Report on the Follow-up of the Second Ministerial Conference.* Ministry of Agriculture and Forestry Liaison Unit, Helsinki.

Ministry of Agriculture and Forestry (1996a) *Intergovernmental Seminar on Criteria and Indicators for Sustainable Forest Management – Background document.* Ministry of Agriculture and Forestry, Helsinki.

Ministry of Agriculture and Forestry (1996b) *Intergovernmental Seminar on Criteria and Indicators for Sustainable Forest Management – Final document.* Ministry of Agriculture and Forestry, Helsinki.

Mitchell, G. (1996) Problems and fundamentals of sustainable development indicators. *Sustainable Development* 4, 1–11.

Montreal Process (1996) http://www.iisd.ca/linkages/forestry/mont.html

Noss, R.F. (1990) Indicators for monitoring biodiversity: a hierarchical approach. *Conservation Biology* 4, 355–364.

OECD (1995) *OECD Core Set of Environmental Indicators: Forest Resources.* Working Paper. OECD Group on the State of Environment, Paris.

Ott, W.R. (1978) *Environmental Indices: Theory and Practice.* Ann Arbor Science, Ann Arbor.

Prabhu, R., Colfer, C.J.P., Venkateswarlu, P., Lay Cheng Tan, Rinesko, S. and Wollenberg, E. (1996) *Testing Criteria and Indicators for the Sustainable Management of Forests. Phase 1.* CIFOR, Bugos, Indonesia.

Rees, W. (1994) Revisiting carrying capacity: area-based indicators of sustainability. In: Moser, F. (ed.) *Evaluation Criteria for a Sustainable Economy.* Proceedings EFB Event 90/1994, Institut für Verfahrenstechnik, Technische Universität Graz, Graz, Austria.

Saleth, R. (1994) Social implications and their measurement of sustainability. In: Moser, F. (ed.) *Evaluation Criteria for a Sustainable Economy.* Proceedings EFB Event 90/1994, Institut für Verfahrenstechnik, Technische Universität Graz, Graz, Austria.

UNCSD-IPF (1996) Category III: Scientific research, forest assessment and development of criteria and indicators for sustainable forest management. Paper presented to the Commission on Sustainable Development Ad Hoc Intergovernmental Panel on Forests, Second Session, 11–22 March 1996. E/CN.17/IPF/1996/10. UN Department for Policy Coordination and Sustainable Development.

Attitudinal and Institutional Indicators for Sustainable Agriculture

17

Philip Lowe, Neil Ward and Clive Potter

INTRODUCTION

Regulation of agriculture's environmental impacts is not simply a technical problem, but involves the reorientation of a social and occupational community. Education and efforts to change values and institutional structures must therefore be part and parcel of any effective regulatory strategy to promote a more sustainable agriculture. Whether under the mantle of sustainability or not, action is already being taken to tackle at least some environmental problems, the absence of which would be a feature of a sustainable agriculture. In most developed countries, for example, policies intended to reduce agricultural pollution and protect semi-natural habitats and valued landscapes are already in place and directing agriculture in particular ways.

In general, the regulatory framework will channel the course of agricultural development in so far as it modifies present agricultural practices and induces technical change on the farm. This relationship between regulation, technical change and farm adjustment may be conceived as one of 'structural learning'. Structural learning (Argyris and Schon, 1978) focuses on changes to the cognitive or normative propositions held by individual or collective actors. From this perspective, the process of moving towards a more sustainable agriculture will be marked as much by changes in the interpretive frames of agricultural actors (i.e. developments in their understanding) and related institutional developments (that reflect the new connections and relationships being forged) as it will be by the achievement of more commonly discussed and measurable outcomes, such as reduction in pollution and increases in biodiversity. This shifts the correct target for monitoring effective change from events to the extent of learning processes: genuine change will be reflected in new attitudes and approaches. Yet the focus on environmental *outcomes*, which is reinforced by the preoccupation with quantitative indicators, may lead to a disregard of the lessons being learned from regulatory policy by target groups and the need for wider institutional reforms. It is unclear, for example, whether current regulations are helping to catalyse structural learning in the direction of a more sustainable agriculture or are acting to fortify incompatible behaviours and attitudes. While it is possible to gather evidence to make judgements on precisely these matters, it is also the case that the full environmental consequences of the measures taken in recent years to counter the adverse effects of contemporary agriculture will not become clear for several years to come. Attitudinal and institutional indicators thus allow us to judge in the meantime whether policy is set in an appropriate direction.

In this sense, appropriate attitudinal and institutional developments may be seen as a means of achieving desired environmental outcomes. But such developments may also be seen as desired ends in themselves. Moving towards sustainability is likely to be eased by a shift from tackling specific problems to a focus on agricultural systems and on integrating environmental considerations into them. Indeed, the real challenge – the ultimate objective – is to establish sustainable systems with a capacity, at the individual and institutional levels, for *self-monitoring*. Any specifications of environmental outcomes, in contrast, must be recognized as necessarily provisional and subject to revision as a result of improved scientific understanding, debate over social priorities and the changing state of the environment. This chapter seeks to propose a set of institutional and attitudinal indicators to gauge progress in this fundamental reorientation of agriculture. We begin first by looking at the recent vogue for quantitative environmental indicators and by considering their shortcomings as instruments of policy analysis.

THE DEVELOPMENT AND SHORTCOMINGS OF QUANTITATIVE ENVIRONMENTAL INDICATORS AS INSTRUMENTS OF POLICY ANALYSIS

Nikolas Rose has remarked that, 'modern political argument seems inconceivable without some numerical measure' (1991, p. 674). The appeal of numbers to decision-makers faced with many conflicting and competing demands is that numbers purport to act as 'automatic technical mechanisms for making judgements, prioritising problems and allocating scarce resources' (Rose, 1991, p. 674). Numbers give at least the appearance of objectivity, rendering invisible the social choices their collection and collation entail (see also Porter, 1995).

Information on environmental trends – on air or water quality, on wildlife numbers, on the loss of farmland – has been collected in several countries over many years. Efforts to coordinate, standardize and collate this information between countries began in the 1970s, which is when the European

Commission and the Organisation for Economic Co-operation and Development (OECD), each first issued a state of the environment report. The vogue for environmental indicators of more recent years stems from a number of factors. As environmental concerns rose up national and international agendas throughout the 1980s, increasing demands were made to place environmental considerations on a par with economic objectives in national decision-making. Politicians were used to judging the performance of governments by reference to standard economic indicators: due regard for the environment implied the need for a parallel set of indicators. Environmental problems were recognized to be complex and multi-causal, and the development of a core set of indicators, it was assumed, would at least highlight the major trends and clarify the key relationships, and might also provide early warning of potential problems.

With governments required to pay ever more attention to environmental concerns, there was a need for systematic information to enable officials, politicians and the public to monitor what progress, if any, was being achieved. The expanding array of international agreements in the environmental field also called for transnationally comparable data to help states compare their performance and assess the compliance of others. European environmental policy is a special case in this regard. It represents by far the most elaborate international legal regime for environmental management and regulation, and one that is distinguished by its supernationality and shared sovereignty. Moreover, the specific needs for standardized data for policy-making and compliance monitoring on a European Union (EU)-wide basis have thrown up a separate organization, the European Environment Agency, dedicated to this purpose (Wynne and Waterton, 1998).

Several countries have been developing environmental indicators for some time (e.g. Adriaanse, 1993; Bakkes, 1994). The OECD work on indicators, based on the pressure–state–response concept, has proved influential with other organizations. The premise is that human activities exert *pressures* on the environment that change its *state* (i.e. its quality and its stock of natural resources) and that society *responds* to these changes through shifts in policy and behaviour which, in turn, affect the human *pressures* on the environment.

A number of organizations, including the World Bank (1995) and the World Resources Institute (1995), have proposed modifications to this framework. In particular, the term *driving forces* has been substituted for that of pressures in recognition of the variety of factors, human and non-human, that cause environmental change (OECD, 1997). Work is also going on within the UN Commission on Sustainable Development (1995), EUROSTAT (the Statistical Office of the European Communities) and the European Environment Agency. While much of the earlier work focused primarily on environmental indicators, increasingly countries and international organizations are working to develop indicators of sustainable development, although there is no consensus as yet on what these should cover.

In the United Kingdom (UK), local authorities have developed a set of indicators for use at the local level (Local Government Management Board, 1995). In 1994, the government published its *Strategy for Sustainable Development* (DoE, 1996) which included a commitment to develop a set of appropriate indicators. An interdepartmental Working Group was set up which worked to a framework based on the key issues and objectives set out in the Strategy. Its report put forward a 'preliminary set' (DoE, 1996, p.3) of about 120 indicators, including not only environmental ones but also those explicitly linking environmental impacts with socio-economic activity, in keeping with the sustainable development perspective. In 1998, these were reviewed, as part of the process of revising the *Strategy for Sustainable Development*, in order 'to produce a more comprehensive core set, and also to identify a handful of indicators which will provide an overview of progress. These will need to encapsulate the range of environmental, social and economic issues which make up sustainable development' (DETR, 1998, p.6).

Clearly, quantitative environmental indicators have their uses but they also have their limitations, especially for policy analysis which is their main justification. One of the key limitations is that such indicators express only what is quantifiable. Some environmental characteristics are inherently more amenable than others to quantification. Concentrations of pollutants, for example, can be measured with a reasonable degree of precision, while other characteristics, such as aesthetic beauty or tranquillity or the quality of land management, are recognized to defy objective measurement despite their importance (CPRE, 1995). Rather than difficult issues that should be set to one side, these matters are at the nub of the environmental critique of forms of economic development whose preoccupation with quantitative expressions of growth overrides considerations of the quality of life. Yet the emphasis on measurable outcomes leads to a pernicious form of issue redefinition and displacement that marginalizes such matters. For example, biodiversity indicators are proposed as surrogates for landscape quality even though the latter is not reducible to the former (there are many pleasing landscapes that are ecologically unexceptional and many visually unappealing mudflats teeming with wildlife).

Reliance on quantitative indicators may thus lead to misrepresentation of the issues and the distortion of priorities. The risk is that action will be concentrated where it will particularly influence the available indicators – a problem common to performance measures of any type. Alternatively, governments may be tempted to set objectives that are measurable but not particularly meaningful – for example, pesticide reduction programmes expressed in terms of the overall mass of active ingredients rather than the environmental risks they pose.

Information is most often sought to explain why things are changing over time or to interpret the effect of particular governmental or societal actions on the environment. Unfortunately, another shortcoming of indicators is that causation cannot necessarily be inferred from a correlation between the behaviour of particular indicators – that would necessitate detailed investigation. Indicators reveal trends: they do not explain them.

Another difficulty lies in developing indicators specifically for *sustainable development*. This necessitates going beyond measures simply of environmental protection or degradation to indicators that explicitly link impacts with socio-economic activity. However, such linkages are rarely straightforward and great caution or insight is required in interpreting them. The shift in emphasis towards sustainable development should highlight processes and institutional structures rather than specific environmental outcomes, the task being to establish sustainable systems rather than to achieve some environmental end-state.

A final shortcoming of quantitative indicators relates to the problem of time. The significance of most environmental indicators is what they reveal in the medium to long term, typically over several years, if not decades, and they are usually tracked on an annual basis or longer. Short-term fluctuations – say in wildlife populations or nutrient balances – may not be significant and may simply be due to variable weather conditions, for example. The effects of government policies and other societal interventions on the environment usually also have a long time lag. The consequence is that *current* trends are unlikely to be a reliable indication of the performance of *current* policies. Indicators may therefore be a guide to problems that need to be addressed but are of little help in assessing the policy response. The claim that indicators 'can help to measure the extent to which policies aimed at sustainable development objectives are being achieved' (DoE, 1996, p.2) is, therefore, something of a hollow one.

Underlying these specific shortcomings is a flawed analogy with economic statistics. The notion that the use of environmental indicators should mirror the role of economic indicators in the transactions of government implies that the issue is essentially one of management, of minor adjustments to the pace or direction of development – just as signs of an upward trend in inflation or unemployment might call for a slight tightening of monetary policy or a slackening of fiscal policy. But it is naïve to imply that the only reason policies and practices in the past were unsustainable was because of lack of refined information concerning the consequences. More typically, it was because these policies and practices were conceptually flawed in ignoring the environment as a factor in their models of action and intervention. The shift to a sustainable trajectory must therefore involve the redesign of policies, institutions and structures. To do this purposively calls for causal knowledge, not just trend knowledge of processes. And that necessarily entails disaggregated, sector-specific conceptual models, not the sort of aggregated societal thinking that lies behind the OECD's indicator framework (with its echoes of macroeconomic models and its search for 'core' indicators).

MONITORING AGRI-ENVIRONMENT POLICY

The points made above apply particularly to the agricultural sector. Interest in the development of agri-environmental indicators is growing, driven mainly by a concern on the part of policy-makers to be seen to be demonstrating the environmental effectiveness and value for money of the new wave of agri-environmental schemes. According to an evolutionary model of agricultural policy change, the 'decoupling' process, once begun, puts mounting pressure on policy-makers to find new ways of justifying agricultural support. This is because substituting direct payments for market support increases the visibility and transparency of government transfers, shifting the burden from 'rationally ignorant' consumers to taxpayers, and engendering a debate about the extent to which farmers deserve to be supported *simply because they are farmers*. Although paying farmers to protect the environment is, in principle, more publicly defensible in these terms, such a policy shift, according to some commentators, brings with it a need to develop methodologies that will measure the environmental return on every ECU spent. More specifically, there are pressures on governments to be able to justify their support of farmers to other governments and to do so within a neo-liberal ideology. This would seem to require the demonstration of commensurable public benefits devoid of market-distorting effects.

The debate about agri-environmental policy is thus entering a new phase, with policy-makers becoming increasingly anxious to measure, not only policy outputs – the setting up of the schemes themselves and the enrolment of farmers and land – but also the outcomes – the long-run environmental changes actually produced on farms and in fields. But describing the former is a good deal easier than assessing the latter, and having (apparently) demonstrated that farmers are willing and able to take up agri-environmental schemes in sufficient numbers potentially to make a difference, agriculture departments are facing increasing demands for evidence of the 'additionality' effects of schemes: the extent to which they succeed in bringing about changes in farming practices that would not otherwise have occurred, and the public value of the results.

Having said that, monitoring and assessment studies have tended to concentrate on describing the rate of uptake of schemes and their farm income effects, largely skirting round the methodological problems associated with measuring additionality effects. For example, the evidence so far regarding Environmentally Sensitive Areas (ESAs) in the UK suggests that, in socio-economic terms, there have been substantial additions to farm income, but that changes in the management of land on the farms sampled have been slight. In the UK, ESAs are now a fairly mature policy measure. The first stage of designations was between 1987 and 1989. This scheme has also been closely monitored and assessed, perhaps more so than for any other national agri-environment measure in Europe. The majority of land has been entered by farmers at the lowest tiers, requiring participants in effect to maintain existing practices in order to qualify for payment. Ecological monitoring confirms that very little change in landscape or botanical terms has taken place on ESA land during the first few years of the policy's operation.

The conclusion of an overview of the monitoring and evaluation studies done to date was that 'For the time being there is little detectable change in habitats or landscapes which can be attributed to the ESAs' (Whitby, 1994, p. 264). Nevertheless, ESAs have been judged a success. For example, they enjoy strong backing from environmental groups who approve of this form of support to farmers and who understand that any environmental benefits will be realized in the medium to long term (House of Commons Agriculture Committee, 1997). Such interim judgements need to be made in default of conclusive evidence of environmental results.

More generally, we must accept that the full environmental consequences of many of the measures taken in recent years to counter the adverse effects of contemporary agriculture will not become clear for several years to come. For a start, while it is possible to refer to the take-up rates of various schemes and changes in farming practices, the consequent second-order effects in such areas as changes in enterprises, farming systems or technology and farm structure may be complex and protracted (Lowe *et al.*, 1990). Secondly, the response of the natural environment may be particularly prolonged. For example, increasing the biodiversity of agriculturally improved grassland may take up to 100 years; and to attenuate harmful levels of soluble phosphorus in soils may take up to 50 years. Policy-makers must be aware of this. To anticipate that the agri-environment schemes introduced following the 1992 reform of the Common Agricultural Policy should generate significant and tangible (and therefore measurable) environmental results in time, say, for the reopening of world trade talks on agriculture in 1999 is simply to raise false expectations, whatever the objective merits of the schemes.

How then are we to judge the potential appropriateness of current initiatives? Implicitly or explicitly, policy interventions make assumptions about causality and agency. The regulation of pesticides largely at the point of registration rather than the point of use is one example; the regulation of farm waste at the point of disposal is a contrasting example (see Clark *et al.*, 1994; Ward et al., 1998). But are the assumptions about causality and agency embedded in policies in keeping with the nature of the problem and the requirements of sustainable development?

AGENCY AND CAUSALITY IN AGRI-ENVIRONMENTAL CHANGE

The emergence of an array of social and environmental problems associated with the technological transformation of agriculture has required the reinstatement of human agency into our understanding of agricultural change (Lowe *et al.*, 1992). On the one hand, models of technological determinism – the inevitable, inexorable and unilinear progression of technological change –

have been challenged. On the other hand, campaigners and law-makers have sought to attach responsibility for the various environmental problems that have arisen as a result of agricultural change.

Allocating responsibility is not just a matter of apportioning blame but also of understanding where and how to intervene to address a problem. As particular environmental problems have arisen in relation to agriculture, so the regulatory responses have targeted specific agents in the agro-food system, thus assuming a certain pattern of causal links within the system. The technology available is a crucial factor in both determining the requirements for regulation and setting limits on the regulatory standards which realistically can be imposed. In short, regulation and technical change coevolve and the two together will foreclose some options for future farming practice while augmenting the chances of other possibilities.

Critics of models of technological determinism have revealed the way successful technological paradigms depend upon the establishment and maintenance of social and institutional networks. Such networks for agriculture typically link together research and development (R&D) institutions, manufacturers, suppliers, advisors and producers, and are supported by the regulatory framework. The arrangements and principles thereby forged to promote and sustain past technological developments also set the preconditions for successor products and practices which, if compatible, flow with ever greater ease along the established networks (Dosi, 1982; Hughes, 1987; Ward, 1995). Farmers are thus embedded in complex techno-economic networks that encourage the development of agriculture along a particular trajectory. The momentum generated by previous rounds of technological change can prove a significant barrier to redirecting that trajectory. The corollary is that the effective promotion of alternative technological paradigms will require the establishment of new networks and changes to the policy and regulatory framework, as well as the redirection of R&D and farm advisory services (Ward, 1996; Winter, 1997).

It was argued above that the relationship between regulation, technical change and farm adjustment should be conceived as one of 'structural learning' (Argyris and Schon, 1978). The notion of structural learning focuses on change to the cognitive or normative propositions held by individual or collective actors. Learning is thus not equated with the achievement of specific targets, such as the accomplishment of particular tasks, but with the change of interpretive frames held by actors. New interpretive relationships between facts, events, ideas and so on can (but do not have to) lead to changes in behaviour. Thus structural learning is not the changed behaviours of actors (i.e. a dependent variable) but the change in interpretive frames (i.e. an independent or intervening variable) as expressed in new attitudes and approaches. This raises questions about the reasons for effective or ineffective learning, for impediments to transforming interpretive frames into action.

Sustainability requires the integration of environmental objectives into economic activity at all levels (in terms of changed interpretive frames this implies that actors should make the connections to the environmental implications of their actions and decisions). Ostensibly, this is the main purpose behind the intergovernmental efforts to devise environmental indicators, or as the OECD puts it, 'for the integration of environmental and economic decision-making, at national and international level' (OECD, 1991, p. 8). However, the macro and managerialist perspective adopted obscures the social and political forces at work. Thus, on the one hand, in conceiving the relation between agriculture and the rural environment in terms of technical adjustments to an economic sector (rather than the reorientation of a social and occupational community), the OECD-type indicator framework perpetuates a policy outlook which, by abstracting farming from its social and environmental context, is part of the fundamental problem. On the other hand, in seeking to construct an objective, politically neutral indicator framework, the OECD approach deliberately obfuscates the political conflicts involved, in which national governments and powerful sectional groups have strong vested interests. The consequence is a failure to acknowledge that effective integration must imply changed social relationships and institutional structures.

To analyse the response of agriculture in a way that centrally addresses the social and political dimensions of integration, we can usefully distinguish the following levels: the individual farmer; the regulation of farming (i.e. the pattern of incentives and controls acting on farmers); and policy-making for agriculture. Below, for each of these levels, we consider in turn the sorts of indicators that would reveal significant adaptive responses integrating environmental objectives.

INTEGRATION OF ENVIRONMENTAL OBJECTIVES INTO FARMERS' ATTITUDES

Agricultural policy reforms in general, and agri-environment measures in particular, should enhance farmers' attitudes towards the environment and clarify their understanding of their environmental responsibilities. It might reasonably be expected that there would already be discernible changes in farmers' attitudes, and even in farming cultures, from participation in agri-environment schemes, even where the environmental consequence could still not be gauged. Previous research (see, for example, Potter and Gasson, 1988) suggests that the attitude of farmers entering agreements is a critical determinant of the level and quality of any environmental benefits obtained. Colman *et al.* (1992) have argued that 'policy measures which encourage positive attitudes to conservation will in the long term be more effective than those that do not, since a positive shift in attitudes will increase the output of conservation goods at any specified level of budgetary cost'. Indeed, it could be argued that unless they exert such an influence, agri-environmental schemes will inevitably be seen as temporary bribes, shallow in operation and transitory in their effect (Morris and Potter, 1995).

Recording changes in the attitude and outlook of farmers participating in agri-environmental schemes may be used as an indicator of farm-level structural learning. Unfortunately, no research has yet investigated the extent to which environmental attitudes evolve during the course of a management agreement. Work already conducted in the UK, however, has analysed the motives of farmers entering agreements and has sought to measure any differences in outlook and situation compared to non-participants. This more behavioural approach has provided some clues about the depth of commitment to and engagement with programme objectives, and the likelihood that conservation activity will be sustained in the long term.

Farmers can be placed on a 'participation spectrum', ranging from the most resistant non-adopters through to 'compliers' (farmers who conform to the terms of management agreements in order to receive the payment), and concluding with a small number of self-selecting 'stewards' (farmers who evince a strong conservationist mind set). There is little evidence of any movement between these categories as a result of initial participation in schemes. Comparative research in the UK, The Netherlands, Germany, France and Spain (Lobley and Potter, 1998) suggests that the bulk of current participants in most agri-environmental schemes within the EU are effectively 'compliers' on this definition, joining schemes because of the goodness of fit between scheme design and their existing farming system. Stewards are very definitely in the minority, their participation typically being a continuation of a long trajectory of environmental management on the farms concerned.

In additionality terms, policy-makers would seem to be caught in a double bind. On the one hand, scheme conditions need to be sufficiently undemanding to attract enough farmers into schemes to make a difference. This may be environmentally useful to the extent that existing features and habitat mosaics are maintained, but additionality effects are likely to be difficult to prove empirically. It also encourages compliance behaviour on the part of farmers which can easily be abandoned once contracts expire. On the other hand, more restrictive (usually upper-tier) agreements, necessary to engineer more substantial changes in land use and the restoration of landscapes and habitats, appear to be appealing to a much smaller, self-selecting band of farmers. Because many of these would have done what they are doing without subsidy, through force of circumstance or conviction, it is a moot point whether they are generating any more additionality than the compliant majority. However, being much more genuinely engaged with the environmental objectives of schemes, the endorsement of the values of these participants and any additions to their ranks must be counted as achievements in the long-term objective of moving towards a sustainable agriculture.

Training of farmers may be crucial if attitudes are to be changed more widely. Without appropriate training, the impact of schemes on farmers' attitudes and behaviour may be shallow and temporary. Participation in agri-environmental training programmes is therefore likely to be a better indicator of progress than participation in agri-environmental schemes. It is important that training programmes pay as much attention to structural learning as to content learning (i.e. the why and not just the how of conservation).

INTEGRATION OF ENVIRONMENTAL OBJECTIVES INTO THE REGULATION OF FARMING

To achieve sustainability involves not only the encouragement of environmentally sensitive attitudes and practices, but also the assumption by the farmer of the role of environmental manager. The progress of integrating environmental objectives into the regulation of farming should be judged in terms of the extent to which it casts farmers in this role. Of central importance is a change in the relationship between technological change and farming practice. Post-war agricultural policy adhered to a model of agricultural innovation in which the dynamic of technological advance was seen to lie outside of farming, in the laboratories of the supply companies, the universities and the agricultural research institutes. Farmers were seen to have little influence over the process or its consequences, other than the rate at which they chose to take up the technologies available. But even in this regard, competitive pressures made them very susceptible to new techniques that lowered production costs and enhanced productivity. The term widely used to characterize this dependency was the 'technological treadmill' (Ward, 1993).

The role of farmers as environmental managers emphasizes a different model of technological change – that of farmers as *adapters* and not simply adopters of available technology. Thus, whereas a key role of technology in the past has been to overcome and eliminate environmental variability and constraints, now conversely the emphasis is on farmers carefully adapting technology to respect environmental variability. The farmers' operational knowledge therefore must not just be derivative of that of the agricultural scientists and technologists, but must also draw upon an intimate understanding of the farm environment and its physical, ecological and meteorological variability. Farm-based strategies for environmental management must thus combine scientific and indigenous knowledge (Murdoch and Clark, 1994).

There are various formal methods for integrating the environment into farm management. These include farm waste management, farm conservation and farm sustainability plans and codes of good agricultural practice. One indication of progress would be the extent to which farmers had taken up and implemented such methods. Conservation advice obviously has an important role to play and is being incorporated into farm advisory and extension services. This can be done to a token or marginal extent, or it can be given a more central role – possible indicators would be the proportion of farm advice devoted to conservation and environmental protection and the extent of the availability of such advice (Mitchell and Baldock, 1996). At the same time, the vertical integration of farmers needs to be counteracted to reduce their technological dependency and, instead, horizontal networks promoted that expand the self-monitoring and learning capacity of farmers in relation to their local environment (Winter, 1997). Examples would include farmer self-help groups orientated towards conservation and farmer–environmentalist networks. Another indicator would be the spread of such groups.

INTEGRATION OF ENVIRONMENTAL OBJECTIVES INTO AGRICULTURAL POLICY-MAKING

Environmental issues and concerns represent a challenge for traditional functions of government that are organized along sectoral lines. The gathering debate over sustainable development led in the 1980s to a new emphasis on seeking to establish a more synoptic environmental policy with coordinated environmental goals integrated into each sector. Potentially, this had a number of dimensions: at the strategic level – a global approach to the setting and monitoring of environmental objectives; at the sectoral level – an emphasis on integrating environmental goals into sectoral objectives (e.g. the greening of transport, tourism, etc.); at the level of policy instruments – use of instruments such as cross-compliance, economic mechanisms and environmental assessment procedures.

The commonly identified barriers to integration involve both a political and an organizational aspect. In the case of the former, a traditional problem for environmental policy has been its relatively low priority on the political agenda, reflecting the dominant attachment of governments to the imperative of economic growth. Environmental protection has generally been treated as something of a luxury to be afforded in times of economic prosperity, or to be traded off against material goals. Moreover, environmental problems are often deep-seated and many environmental policies only promise benefits in the long term; thus, they are often ignored or sacrificed in political systems geared up for short-term electoral or economic cycles. Sectoral economic policies, on the other hand, tend to be supported by strong producer organizations whose members are usually directly and significantly affected by government action, in contrast to the diffuse public benefits yielded by environmental policy. In consequence, environmental policy may lack the weight to compete for resources or challenge the policy assumptions of other issue arenas and can be marginalized within institutional structures.

In terms of organizational barriers, governmental bureaucracies tend to be highly compartmentalized, which means they are not well designed to absorb cross-cutting environmental concerns. The approach to integration adopted will depend upon the degree of formal power or authority at the disposal of the environment ministry. Among EU member states, most environment ministries have been established only relatively recently: the oldest have been in existence for less than 25 years and some (as in Italy and Spain) for less than 10 (Wilkinson, 1997). Traditions of policy-making and styles of regulation differ between sectors and with the approach adopted in environmental policy. The integration of environmental objectives is therefore obviously not an overnight process. There is likely to be a different pace between states and some policy sectors will respond more readily than others.

In most European countries, agriculture has traditionally been a particularly closed policy community, embracing agriculture ministries and mainstream farming lobbies but to the exclusion of other interests. For a long time agricultural policy communities resisted both the imposition of environmental constraints and the incursion of environmental interests. The acceptance of agri-environmental schemes and measures therefore represents a positive departure but does not necessarily imply integration as much as the defensive co-option of certain environmental issues and values by the agricultural policy community.

The devising and implementation of effective agri-environment policies, however, calls for the involvement of organized environmental interests. Where environmental agencies or lobbies are weak or still remain excluded, there are limited counter pressures to the complete internalization of agri-environment policies within the agricultural bureaucracy which can result in very little public debate or independent evidence about the nature, purpose or achievements of such policies.

Environmental integration must clearly involve institutional change. Liberatore (1997), in attempting to set up a framework for measuring the different degrees of integration across sectors, highlights five major dimensions:

1. Environmental impact assessments – to what extent are these carried out in the sector? Is their use *ex ante* or *ex post*?
2. Consultation – what degree of consultation takes place with authorities having environmental competences at the local, national and EU levels? Does it reflect co-decision making or is it merely symbolic?
3. Compatibility – when producing legislation and regulations for the sector, is compatibility with environmental legislation assessed?
4. Monitoring and evaluation – is any systematic evaluation of the environmental consequences of policies, research or economic activities conducted?
5. Funding – is funding made available for environmentally friendly actions in the sector and on what scale?

Coverage of each of these dimensions in depth and on a regular basis would imply full-scale integration. However, when they are pursued on an *ad hoc* and limited basis then environmental integration is much diluted, and may indeed have little impact on certain sectors.

Hey (1997) distinguishes three types of approaches to integration that different sectors have adopted: defensive, indirect and active. *Defensive integration* attempts to contain and offset possible environmental side-effects arising from the policies already being pursued in the sector. *Indirect integration* arises where existing sectoral policies do give rise to positive environmental benefits, but largely as unintended side-effects. *Active integration* occurs where planned environmental targets, objectives and policy instruments are adopted within a sector. Such embedding of environmental integration into sectoral decision-making processes is likely to require modifications to organizational structure. So far, it would seem, the integration of environmental objectives into agricultural policy has involved varying degrees of defensive and indirect integration.

CONCLUSION

The effective integration of environmental objectives into agriculture entails changed social relationships and institutional structures linked to altered interpretive frames on the part of those responsible. While it is possible now to gather systematic evidence of the extent to which these changes towards a sustainable system are taking place, it will be some years before the specific environmental outcomes of recent policy developments can be judged. The current preoccupation with physical environmental indicators is therefore somewhat misplaced. It represents a typical case of means–ends displacement. After all, it is permanently sustainable socio-economic systems for which we should be striving, not some sort of environmental end-state. But the need for social and institutional change always makes powerful groups and technocrats uneasy.

REFERENCES

Adriaanse, A. (1993) *Environmental Policy Performance Indicators – a Study on the Development of Indicators for Environmental Policy in the Netherlands*. Ministry of Housing, Physical Planning and the Environment, The Hague.

Argyris, C. and Schon, D. (1978) *Organisational Learning*. Addison Wesley, London.

Bakkes, J. (1994) Research needs. In: Moldan, B. and Billharz, S. (eds) *Sustainability Indicators*. John Wiley and Sons, Chichester.

Clark, J., Lowe, P., Seymour, S. and Ward, N. (1994) *Sustainable Agriculture and Pollution Regulation in the UK*. Centre for Rural Economy Working Paper Series No. 13, University of Newcastle upon Tyne.

Colman, D., Crabtree, R., Froud, J. and O'Carroll, L. (1992) *Comparative Effectiveness of Conservation Mechanisms*. Department of Agricultural Economics, University of Manchester.

CPRE (1995) *Measuring the Unmeasurable*. Council for the Protection of Rural England, London.

DETR (1998) *Opportunities for Change: Consultation Paper on a Revised UK Strategy for Sustainable Development*. Department of the Environment, Transport and Regions, London.

DoE (1996) *Indicators for Sustainable Development for the United Kingdom*. HMSO, London.

Dosi, G. (1982) Technological paradigms and technological trajectories: a suggested interpretation of the determinants and directions of technical change. *Research Policy* 11, 147–162.

Hey, C. (1997) Greening other policies: the case of freight transport. In: Liefferink, D. and Skou Andersen, M. (eds), *Innovation in European Environmental Policy*. Scandinavian University Press, Copenhagen.

House of Commons Agriculture Committee (1997) *Environmentally Sensitive Areas and Other Schemes under the Agri-Environment Regulation*. HC Paper 45-I. HMSO, London.

Hughes, T. (1987) The evolution of large technological systems. In: Bijker,W., Hughes, T. and Pinch, T. (eds) *The Social Construction of Technological Systems: New Directions in the Sociology and History of Technology*. MIT Press, London, pp. 51–82.

Liberatore, A. (1997) The integration of sustainable development objectives into EU policymaking. In: Baker, S., Kousis, M., Richardson, D. and Young, S. (eds) *The Politics of Sustainable Development: Theory, Policy and Practice within the European Union*. Routledge, London, pp. 107–126.

Lobley, M. and Potter, C. (1988) Environmental Stewardship in Britain's Countryside. *Geoforum*, in press.

Local Government Management Board (1995) *Sustainability Indicators*. LGMB, Luton.

Lowe, P., Marsden, T. and Whatmore, S. (eds) (1990) *Technological Change and the Rural Environment*. Fulton, London.

Lowe, P., Ward, N. and Munton, R. (1992) Social analysis of land use change: the role of the farmer. In: Whitby, M. (ed.) *Land Use Change: Causes and Consequences*. HMSO, London, pp. 42–51.

Mitchell, K. and Baldock, D. (1996) *Farm conservation advisory services in selected Member States*. Institute for European Environmental Policy, London.

Morris, C. and Potter, C. (1995) Recruiting the new conservationists. *Journal of Rural Studies* 11, 51–63.

Murdoch, J. and Clark, J. (1994) *Sustainable Knowledge*. Centre for Rural Economy Working Paper No. 9, University of Newcastle upon Tyne.

OECD (1991) *Environmental Indicators: A Preliminary Set*. Committee for Agriculture, Organisation for Economic Co-operation and Development, Paris.

OECD (1994) *Environmental Indicators*: OECD Core Set. Committee for Agriculture, Organisation for Economic Co-operation and Development, Paris.

OECD (1996) Developing OECD *Agri-Environmental Indicators*. Committee for Agriculture, Organisation for Economic Co-operation and Development, Paris.

OECD (1997) *Environmental Indicators for Agriculture*. Organisation for Economic Co-operation and Development, Paris.

Porter, T. (1995) *Trust in Numbers: The Pursuit of Objectivity in Science and Public Life*. Princeton University Press, Princeton, New Jersey.

Potter, C. and Gasson, R. (1988) Farmer participation in voluntary land diversion schemes. *Journal of Rural Studies* 4, 365–375.

Rose, N. (1991) Governing by numbers: figuring out democracy. *Accounting, Organization and Society* 16, 673–692.

United Nations (1995) *Chapter 40: Information for Decision-making and Earthwatch*. Economic and Social Council, Commission on Sustainable Development E/CN.17/1995/7. United Nations, New York.

Ward, N. (1993) The agricultural treadmill and the rural environment in the post-productivist era. *Sociologia Ruralis* 33, 348–364.

Ward, N. (1995) Technological change and the regulation of pollution from agricultural pesticides. *Geoforum* 26, 19–33.

Ward, N. (1996) Pesticides, pollution and sustainability. In: Allanson, P. and Whitby, M. (eds) *The Rural Economy and the British Countryside.* Earthscan, London, pp. 40–61.

Ward, N., Clark, J., Lowe, P. and Seymour, S. (1988) Keeping matter in its place: Pollution regulation and the reconfiguring of farmers and farming. *Environment and Planning* A, in press.

Whitby, M. (ed.) (1994) *Incentives for Countryside Management,* CAB International, Wallingford, UK.

Wilkinson, D. (1997) Towards sustainability in the European Union? Steps within the European Commission towards integrating the environment into other EU policy sectors. *Environmental Politics* 6, 153–173.

Winter, M. (1997) New policies and new skills: agricultural change and technology transfer. *Sociologia Ruralis* 37, 363–81

World Bank (1995) *Monitoring Environmental Progress. A Report on Work in Progress.* Environment Department, World Bank, Washington DC.

World Resources Institute (1995) *Environmental Indicators. A Systematic Approach to Measuring and Reporting on Environmental Policy Performance in the Context of Sustainable Development.* World Resources Institute, Washington DC.

Wynne, B. and Waterton, C. (1998) Public information on the environment: the role of the European Environment Agency. In: Lowe, P. and Ward, N. (eds) *British Environmental Policy and Europe.* Routledge, London, pp. 119–137.

Discussion and Conclusions

18

Bob Crabtree and Floor Brouwer

INTRODUCTION

The preceding chapters have given a 'state of the art' analysis of indicators, designed to provide information about critical elements in the relationship between agriculture and the environment. The key role played by indicators is to provide environmental information in a summarized form that facilitates communication with users. Recent interest in designing and implementing indicators for the agri-environment reflects a number of pressures facing governments and public agencies. These are:

- public concern throughout Europe regarding the environmental impacts of agriculture and the lack of an adequate and transparent system of measuring and communicating information on environmental change;
- the absence of well-developed procedures for monitoring and evaluating the impacts of agri-environmental policy measures in the context of making such measures more effective and publicly accountable;
- the urgency to establish data sets that are consistently defined and collected so as to allow international comparisons and support the design and management of common policies; and
- the requirements under the next WTO round for quantitative measures of the public benefits associated with environmental (and other) payments to farmers.

Of these, it is the first that has provided the main thrust in public policy for the development of indicators. Governments at national and international level are concerned to produce a limited number of environmental indicators that will be unambiguous, transparent and resonate with the public.

©CAB INTERNATIONAL 1999. *Environmental Indicators and Agricultural Policy* (eds F.M. Brouwer and J.R. Crabtree)

The main theme in this text has been the development and selection of indicators for the agri-environment. From the various contributions we can draw a number of conclusions about indicators and their role in providing information for policy design and evaluation:

- Indicators cannot be developed without a clear context and purpose, in terms of the information to be transferred and the types of target users.
- Whilst indicators must have scientific validity they must also satisfy criteria linked to precision, cost, transparency and communication. In addition, the ability of an indicator to satisfy verification and audit requirements can be critical where they are used as policy instruments, for example in regulation or incentive arrangements.
- There are important questions of scale, time and context. All indicators require a context for their interpretation and this will depend on the appropriate scale; this may be determined by biophysical or administrative considerations. Different indicators of systems performance are usually required at different hierarchical levels (Gallopin, 1997). Thus indicators useful for international comparisons will differ from those for monitoring regional or local environmental change. An appreciation of the time context of an indicator is also vital. For example, where ecosystems are recovering from damage there may be a long time delay between a change observed in a driving force indicator and the response in a related state indicator.
- Pressure and state indicators both have a role and this will differ with context and purpose. State indicators can be more costly to produce, but have the advantage of being more closely linked to environmental endpoints which are the prime public and policy interest. Pressure indicators potentially give a more rapid indication of future environmental states.
- There is a need to explore the role of process indicators both as pressure–state intermediaries and as indicators that measure change in the actors and institutions associated with environmental management and the construction of policy. There is much to be gained from further improving the understanding of links between pressure, process and state indicators, and one mechanism for achieving this is in integrated environmental assessment.
- By summarizing information and concentrating on the measurable, indicators can both distort information and lead to issue displacement. This is also true where indicators are used as instruments in regulatory or incentive policies. Understanding how a set of indicators fail in performance is thus essential to complement the direct interpretation of the indicators themselves.

In examining the linkage between indicators and context, as well as examining the process whereby indicators might be designed and selected, four aspects have dominated the analysis. These are the environmental context, the institutional context, the selection mechanism and the comparison of different types of indicators. This concluding chapter examines these four aspects and highlights some of the unresolved issues that contribute to the research agenda of the future.

ENVIRONMENTAL CONTEXT

Indicators provide information on environmental issues – that is, concerns linked to environmental damage, protection, enhancement and the risks associated with human health. The 'values' associated with such environmental change are of course heavily prescribed by expert opinion and by the activities of environmental interest groups. In broad terms it is now clear which key environmental issues are driving policy in Europe. Impacts on environmental capacity encompass concerns with pesticide, gas and mineral emissions and their impacts on soil, water and air quality. Impacts on the stock of natural assets encompass both negative and positive impacts of agriculture on biodiversity, ecosystems and landscape, and the services that these assets provide (CEC, 1997).

These environmental issues are spatially defined and appear at varying scales. In contrast with the international development of environmental indicators, at regional and local levels there is a demand for indicators that relate to local environmental issues. Sectoral concerns may also suggest scales of measurement (Whitby and Adger, 1996). In the agricultural context, much of the concern to develop indicators is driven more specifically by the needs of policy design and evaluation (Whitby *et al.*, 1998), but scale also has to be selected as appropriate to the environmental target. Global targets (e.g. greenhouse gas emissions) require globally defined indicators that are linked to national emissions. The appropriate spatial scale for water pollution is most evidently the water basin, although this is interlinked both to marine pollution and to specific sub-basin impacts. With biodiversity and landscape, appropriate scales have proved more elusive to define (Wascher, Chapter 6).

INSTITUTIONAL CONTEXT

Issues of scale and location are not confined to the biophysical parameters of the environmental context. As Lowe points out (Chapter 17) indicators are prescribed in an institutional and social context. Since most farmers and landowners have no private interest in recording environmental indicators, some administrative structure has to lie behind the collection of agri-environmental information. Particular problems arise where the administrative scale and boundaries differ from appropriate biophysical scales. This particularly applies where local government seeks to provide environmental indicators (such as under Agenda 21) but where the indicators cannot be readily interpreted at a local or regional scale (Tapper, 1996). Much the same problem applies to the interpretation of indicators for policy monitoring where the biophysical context has a scale which exceeds that of the policy; examples here are landscape, marine pollution and air pollution.

Institutional and stakeholder involvement in indicators raises a number of other important issues. There is a case for developing indicators, however imperfect, as a way of stimulating a debate with stakeholders that contributes to the feedback process of developing improved indicators. Jesinghaus (Chapter 4) presents the argument that progress may best be made by not waiting for the elusive goal of perfect information. National statistical services have an important role here in providing a framework in which data can be consistently recorded and disseminated. In addition, institutional indicators can themselves be set up to measure change in institutions and actors associated with the agri-environment. In a sense this can be considered as part of the process of environmental change and the changing role of policy in delivering environmental improvement. The wider role of 'process' indicators is developed below, but stakeholders and institutions can clearly use indicator selection as a tool to serve their interests in the policy process.

Finally, over time any indicator may become distorted, as the indicator itself, rather than the 'complete' information set, drives a response from actors. This can result in a focus on the changes in the indicators rather than the underlying issue. However, whilst the risk of issue displacement is hardly an argument for not using indicators, it is especially relevant where imperfect indicators are used to guide significant policy decisions.

SELECTION MECHANISMS

The selection of indicators will always rely on scientific and technical analysis. Indeed, of the indicators proposed by the various authors, the main criterion has been technical judgement about the merits of the indicator in relation to existing data availability and the cost and feasibility of developing new measurements. Romstad (Chapter 2) places this more formally in the context of information economics. This provides a decision framework for assessing the benefits from additional indicators (or enhanced indicator quality), the criterion being the value of the information in relation to measurement costs. Despite the problems of assessing value (the multiple uses of many indicators, their spatial and temporal definition, and errors in estimation) this is a theoretical framework that deserves wider application. Little research has been done to quantify the economic benefits from additional precision through choice of more rigorous indicators or greater sampling intensity. What is perhaps surprising is that indicators that are self-evidently imperfect (such as pesticide use as an indicator of loss of biodiversity) may be very informative at relatively low cost.

In some case there will be significant time differences in the information provided. For example, given the slow speed of some ecological responses, a system based on state indicator assessment may take some years to identify significant change (Jakeman *et al.*, 1993). In contrast, indicators that measure ecosystem pressure may pick up expected change much earlier. Irreversible damage is particularly significant in this context. Ultimately what is needed is an application of information economics that allows evaluation of indicators with different time, risk and cost profiles. This would seem to be an important area for the development of efficient systems of monitoring for the agri-environment.

TYPES OF INDICATOR

Most of the contributors used the OECD pressure/driving force–state–response framework as the point of departure for a classification of indicator types. Rather different conclusions have on occasion been reached about the merits of state and pressure indicators (Peco *et al.*, Chapter 10; Tucker, Chapter 7; Van Wenum *et al.*, Chapter 8). The key differences here are that state indicators of ecosystems are typically costly to implement and subject to substantial errors. When evaluating policy they fail to differentiate between policy effects and exogenous-to-policy effects (e.g. climate change). Their merit is that they get close to, or measure directly, the target environmental asset. Pressure indicators are more remote from the change in state but identify potential impacts earlier and often at lower cost. However, these indicators are not mutually exclusive and the cost–benefit of individual contexts needs to be explored in the selection process.

Perhaps more important in its omission from the OECD framework is the idea of process which permeates the indicator issue. For pressure indicators to be useful there must be an implied process model that links pressures to changes in state. It is the ability of the process model to translate changes in pressure into changes in state that underpins the interpretation of the pressure indicator. This opens the possibility of a further set of indicators that measure key processes and go some way towards explaining environmental change. Where direct measurements of changes in the state of the environment are costly or unreliable then process modelling has potential to provide more reliable information about change. Nitrogen balance is an example of such a process indicator.

While process indicators are most obviously biophysical, they also have a role in the institutional context as measures of change in the behaviour of institutions and key agents, such as farmers, in the delivery of policy. Where policy is delivered through intermediate structures, indicators to measure the effectiveness of such processes could be extremely valuable. Recent research on incentive design under imperfect information and on compliance monitoring points the way forward in this sphere (Mirrlees, 1997; National Audit Office, 1997; Latacz-Lohmann, 1998; Moxey *et al.*, 1998). This is an argument for developing so-called integrated environmental assessments that pull together information from a range of indicators. Lowe (Chapter 17) has argued strongly that this assessment needs to be a wide one that incorporates the responses of institutions and target groups in society. These 'response' indicators that measure attitudinal and institutional change are the least well developed of the various types proposed by OECD.

There remain some issues of indicator integration. Oskam (Chapter 11) tackles the question of data redundancy by using principal component analysis to maximize the information content of derived variables. However, other methods for developing composite indicators need to be explored apart from weighted additive systems, and in particular those based on an understanding of the physical processes involved. This integration of indicators could also be developed to define performance indicators by linking context to pressure or state, thus facilitating interpretation. Such performance indicators are especially relevant where critical environmental loads or sustainability thresholds have been defined.

A final point here concentrates on the loss of information inherent in indicators. Since they transmit imperfect information, and this is most apparent if they are composite ones, users need to be aware of the information deficiency when interpreting indicators if distortion and displacement are to be minimized. The use of GDP growth as a proxy for welfare gains (Anderson, 1994) is the best-known example of a universal problem with imperfect information. Hence whilst policy-makers demand simple, reliable indicators that resonate with stakeholders, they need to be aware of the contexts in which a selected set of indicators may distort the development of policy by ignoring key elements for decision-making.

REFERENCES

Anderson, V. (1994) *Alternative Economic Indicators.* Routledge, London.

CEC (1997) *Rural Developments.* DG VI, Commission of the European Communities, Brussels.

Gallopin, G.C. (1997) Indicators and their use: information for decision making. In: Moldan, B. and Billharz, S. (eds) *Sustainability Indicators: a Report on the Project on Indicators in Sustainable Development.* John Wiley & Sons, New York.

Jakeman, A.J., Beck, M.B. and McAleer, M.J. (1993) *Modelling Change in Environmental Systems.* John Wiley & Sons, New York.

Latacz-Lohmann, U. (1998) Moral hazard in agri-environmental schemes. Paper presented at the Agricultural Economics Society Conference held at the University of Reading.

Mirrlees, J.A. (1997) Information and incentives: the economics of carrots and sticks. *The Economic Journal* 107, 1311–1328.

Moxey, A., White, B. and Ozanne, A. (1998) Efficient contract design of agri-environmental policy. *Journal of Agricultural Economics*, in press.

National Audit Office (1997) *Protecting Environmentally Sensitive Areas.* The Stationery Office, London.

Tapper, R. (1996) Global policy, local action? International debate and the role of NGOs. *Local Environment* 1, 119–126.

Whitby, M. and Adger, W.N. (1996) Natural and reproducible capital and the sustainability of land use in the UK. *Journal of Agricultural Economics* 47, 50–65.

Whitby, M., Moxey, A. and Lowe, P. (1998) *Environmental Indicators for a Reformed CAP: Monitoring and Evaluating Policies in Agriculture.* Report to English Nature, Department of Agriculture and Food Marketing, University of Newcastle upon Tyne.

Index

287